Flexible Packaging

A technical guide for narrow- and mid-web converters

Other Labels & Labeling books:

ENCYCLOPEDIA OF LABEL TECHNOLOGY
Michael Fairley

THE HISTORY OF LABELS
Michael Fairley and Tony White

DIGITAL LABEL AND PACKAGE PRINTING
Michael Fairley

ENVIRONMENTAL PERFORMANCE AND SUSTAINABLE LABELING
Michael Fairley and Danielle Jerschefske

CONVENTIONAL LABEL PRINTING PROCESSES
John Morton and Robert Shimmin

LABEL DESIGN AND ORIGINATION
John Morton and Robert Shimmin

LABEL DISPENSING AND APPLICATION TECHNOLOGY
Michael Fairley

CODES AND CODING TECHNOLOGY
Michael Fairley

LABEL EMBELLISHMENTS AND SPECIAL APPLICATIONS
John Morton and Robert Shimmin

BRAND PROTECTION, SECURITY LABELING AND PACKAGING
Jeremy Plimmer

DIE-CUTTING AND TOOLING
Michael Fairley

MANAGEMENT INFORMATION SYSTEMS AND WORKFLOW AUTOMATION
Michael Fairley

SHRINK SLEEVE TECHNOLOGY
Michael Fairley and Séamus Lafferty

LABEL MARKETS AND APPLICATIONS
John Penhallow

PRODUCT DECORATION TECHNOLOGIES
John Morton and Robert Shimmin

INKS, COATINGS AND VARNISHES
Andy Thomas-Emans

ENCYCLOPEDIA OF BRAND PROTECTION
Michael Fairley and Jeremy Plimmer

For the latest list please visit:
www.labelsandlabeling.com

Flexible Packaging

A technical guide for narrow- and mid-web converters

Michael Fairley, LCG, FIP3 and FIOM3
Chris Ellison

Flexible Packaging
A technical guide for narrow- and mid-web converters

First edition published 2018 by:
Tarsus Exhibitions & Publishing Ltd

Second edition published 2020 by:
Tarsus Exhibitions & Publishing Ltd

Printed by Kindle Direct Publishing, an Amazon.com company.

ISBN 978-1-910507-21-6

Contents

While every care has been taken to ensure the information, charts, diagrams and illustrations in this publication are correct at the time of publishing it is possible that technology, specifications, markets and applications, or terminology may change at any time, or that the editor's or contributor's research or interpretation may not be regarded as the latest accepted guidance in some parts of the world of labels.

The publishers therefore cannot accept responsibility for any errors of interpretation or for any actions, decisions or practices that readers may take based on the publication content and would advise that the latest industry supplier specifications, standards, legislative requirements, performance guidelines, practices and methodology should always be sought before any investment or implementation is made.

Preface

Flexible packaging today has one of the highest growth rates across all printing sectors, achieving an annual global growth of close to five percent. Important trends include smaller run lengths, a requirement for multi-versions and variations, personalization and a growing interest in adding in-line value – all undoubted opportunities for narrow- and mid-web label converters (using both flexo and digital technology) to capture a growing portion of the flexibles market.

Jobs with the shortest run sizes are in pouches of any kind, as well as in single serve and one-dose packs and sachets. These are key areas where label converters can most readily enter the flexible packaging market. Indeed, narrower and mid-web conventional and digital printing has effectively opened up the market to a whole range of new customers who have never previously had a solution for short-run flexible packaging orders.

But it's not all quite straightforward for label converters to move into flexible packaging. They need to have a comprehensive understanding of paper, foil and filmic substrates – as well as multi-layer constructions and barrier properties – of inks, coating and curing, of pack sealing methods, and of the specific user requirements needed for the many different types of liquids, powders, gels, creams and solids that have to be packaged. Pre-press requirements are different, inks have to withstand sealing temperatures, web- and sealing widths are important, co-efficient of slip can be critical, while a good understanding of form, fill and seal machine operations is ideal.

This book in the Label Academy series has been designed and written so as to provide both the necessary knowledge and expert guidance that label and other converters looking to enter or expand their production into flexible packaging are most likely to require – a key stepping stone to future success and happy customers.

Michael Fairley
Director, Labels & Labelling Consultancy
Founder, Label Academy

Foreword

Michael Fairley is a beacon in the field of current real-world print education, instructing and inspiring us over the last five years through 'The Label Academy series'. The series meets head on the training needs of the narrow-web label industry, encompassing new market opportunities, materials, printing processes and finishing solutions. Thus, making print and processes, at a time where traditional educational resources are in decline, ever more accessible to new people joining the industry, as well updating and refreshing old hands passionate to stay relevant with the latest technologies.

Mike's latest book, Flexible Packaging - A technical guide for narrow- and mid-web converters (now in its second edition), is an introduction to the world of flexible packaging for both young people at the start of their careers, trainers, supervisors and all those involved in today's busy and evolving converter businesses. Today's converters cannot afford to stand still, technology is ever evolving creating exciting opportunities for printers to push new boundaries and innovate alongside their traditional markets. With innovation comes risk and cost, here Mike's book is an invaluable tool in understanding the basics and building a strong platform of knowledge as converters embark on their own journey.

The book highlights and demystifies the characteristics of flexible packaging, bringing the insight of training experts, OEM's, and the relevance of a converter's approach to the market through compliance and due diligence.

Mike has the gift of making complex concepts accessible to the reader, which for me has made collaborating with him on this book a thoroughly enjoyable experience, raising some questions and testing some theories. Discovering new ideas whilst working on the book has given me an even greater passion for our industry even after 30 years plus as a converter!

I hope the book inspires both passionate entrepreneurs and reignites a desire for education and learning for others about our ever evolving and exciting industry.

Chris Ellison
Managing Director, OPM (Labels and Packaging) Group
President, FINAT

About the Label Academy

This book is part of the recommended study material for the Label Academy, a global training and certification program for the label industry. The Label Academy was created by the team behind Labels & Labeling magazine and the Labelexpo series of events.

The Academy consists of a series of self-study modules, combining free access to relevant articles and videos with paid text books (both printed and electronic). Once a student has completed a module, there is an opportunity to take an online test and earn a certificate.

It is expected that a Label Academy qualification will become a standard in the industry – for printers/converters, suppliers, brand owners and designers – and assist in providing a benchmark. In addition to its own training, the Label Academy will aim to become a resource provider to the many existing educational programs in the industry. Accredited training courses will be promoted through the Label Academy website and books will be provided at discounted rates.

The Label Academy concept was pioneered by industry expert Mike Fairley. This was in response to a reduction in the number of dedicated printing colleges and the need to standardize training across the world. The label industry also has its own specific training needs – it has some of the widest range of materials, printing processes and finishing solutions of any printing sector.

We are also working with other training experts and authors to ensure that the Label Academy provides up-to-date and relevant training material for the industry.

The Label Academy is supported by the key trade associations, including FINAT, TLMI and the LMAI.

www.label-academy.com

Label Academy sponsors

Thank you to our founding sponsors, without whom this ambitious project would not have been possible:

Cerm

Cerm designs business automation software solutions to meet the specific demands of flexo and digital narrow web printers. Using the latest technology, our team's focus is on innovation and continuous improvement.

Our automation solutions support each step in the printer's integrated workflow – from estimating to production, shipment and data collection – and provide the feature and functionality printers need to gain efficiency and improve profitability.

Cerm inspires collaboration and helps printers remain competitive in the market and deliver the best products possible. We are proud to sponsor the Label Academy and contribute to the future of the narrow web printing industry.

www.cerm.net

Flint Group Narrow Web

Flint Group Narrow Web has the products, the solutions, and the technical experts to handle any print situation. Providing solutions for food packaging, sustainability, increased bottom line, efficiency, and uptime – delivering the basics needed to run a successful operation, and the expertise to go above and beyond to another level of success.

Our experts provide solutions to your printing problems with the innovative products and services that have made us an industry leader around the world. Wherever you are, we are available to help you reach your business goals today and into the future.

Continuous improvement is paramount to Flint Group; we are proud to sponsor the Label Academy and the benefits it will bring to the future of our industry.

www.flintgrp.com

Gallus Group

The Gallus Group with its production sites in Switzerland and Germany is a leader in the development, production and sale of narrow-web, reel-fed presses designed for label manufacturers. The machine portfolio is augmented by a broad range of screen printing plates (Gallus Screeny), globally decentralized service operations, and a broad offering of printing accessories and replacement parts. The comprehensive portfolio also includes consulting services provided by label experts in all relevant printing and process engineering tasks. The Gallus Group is a member of the Heidelberg Group and employs around 430 people, of whom 253 are based in Switzerland. The group headquarters is in St.Gallen, Switzerland.

www.gallus-group.com

MPS Systems B.V.

Producing high-quality label printing depends on several factors; one of them is the operator of the press.

As a press machine builder since 1996, MPS Systems B.V. knows how important training and education on subjects like pre-press, label printing and finishing is. For label printers, it is critical that their operators keep up with pre-press and press developments in addition to label trends. Therefore, MPS sponsors the Label Academy, to advance operator's passion for printing, share expertise and help multiply benefits.

The MPS slogans of 'Printers First' and 'Technology with Respect' have always underlined the core philosophy of MPS from press design to operator satisfaction. We develop our presses with a strong focus on user-friendliness and respect for the press operator: Printers First.

www.mps4u.com

HP Indigo

HP Indigo is a global leader in digital printing, with a broad portfolio of digital presses and workflow solutions. Indigo's proprietary Liquid Electrophotography (LEP) technology delivers exceptional print quality for the widest variety of applications including labels, flexible packaging, shrink sleeves and folding cartons. HP Indigo's digital presses match gravure print quality satisfying the most demanding brands.

A division of HP Inc.'s Graphics Solutions Business, Indigo serves customers in more than 122 countries, including many of the top label and packaging converters worldwide.

www.hp.com/go/labelsandpackaging

UPM Raflatac

In a little more than three decades, UPM Raflatac has become one of the world's leading manufacturers of pressure sensitive label materials, developing and leveraging the latest innovations in adhesive technology. Our film and paper label stocks are used for product and information labeling across a wide range of end-uses – from pharmaceuticals and security to food and beverage applications.

We are an engineering driven company with industry-leading products known for their consistent high quality and top performance. We are also known for the high performing supply chain and undisputed leadership in the area of sustainability. UPM Raflatac's dedication to innovation, sustainability and top quality is matched only by our commitment to service excellence. We call it the Raflatouch.

www.upmraflatac.com

About the authors

Michael Fairley
Director, Labels & Labelling Consultancy
Founder, Label Academy

Michael Fairley has been writing and speaking about label and packaging materials, technology and applications since the 1970s, both as the Founder of Labels & Labeling and other print industry magazine titles and as an international consultant writing or contributing to label industry market and technology research reports for the likes of Frost & Sullivan, Economist Intelligence Unit, Pira, InfoTrends and Labels & Labelling Consultancy.

He is the author of the Encyclopedia of Label Technology, co-author of the Encyclopedia of Brand Protection, a contributing author to the Encyclopedia of Packaging Technology and a contributing author to the Encylopedia of Occupational Health and Safety. He also provided significant input to the Academic American Encylopedia.

He now works as a consultant to Tarsus Exhibitions & Publishing – which organizes the Labelexpo shows, Label Summits and publishes Labels & Labeling magazine – as well as regularly speaking at industry conferences and seminars.

He is a Fellow of the Institute of Packaging / Packaging Society, Fellow of IP3 (formerly the Institute of Printing), a Freeman of the Worshipful Company of Stationers, an Honorary Life Member of FINAT and a Licentiate of the City & Guilds of London Institute. He was awarded the R. Stanton Avery Lifetime Achievement Award in 2009.

Chris Ellison
Managing Director, OPM (Labels and Packaging) Group
President, FINAT

Chris Ellison is President of FINAT and Managing Director of UK-based converter OPM (Labels & Packaging) Group Limited, a company established over 40 years ago and today serving more than 600 clients in both domestic and export markets. OPM is a multi-award-winning printer of high quality self-adhesive labels, laminates, sachets, medical and pharmaceutical packaging. Passionate for product innovation, investors in the development of its people and early adopters of total process automation, the company's vision is to remain one of Europe's top performing self-adhesive label and flexible packaging providers.

Acknowledgements

For a label converter looking to diversify and invest in flexible packaging there are many opportunities to develop new products, new markets and new solutions. However, it's not just about utilizing an existing label press or buying a completely new machine; it's also about understanding new materials requirements, coatings and lamination, ink specifications, and knowing what happens to the printed papers, films and laminates during the forming, filling and sealing stages of producing a bag, sachet, pouch, or other type flexible packaging solution.

In putting this book together, extensive reference has been made to relevant master class presentations, to supplier material, including white papers and websites, to looking at the experience of those that have already made the transition. Thanks are due to all those companies that have prepared and made available resource material, documents, press releases and other supporting documents. Their knowledge, help and encouragement has been invaluable.

As with a number of other Label Academy books, Esko have been most helpful in providing illustrative material – this time in terms of images of flexible packaging types and formats from the Esko Flexible Packaging Shapes Store. Their ongoing help and support in building the Label Academy series of books is once again much appreciated.

Thanks are also due to Dr. Marc Heylen, Global R&D and Technical Director, Narrow Web FlintGroup for his technical contribution to the inks and coatings chapter, as well as to MPS Systems B.V. and to Bobst for their input on the printing and converting of flexible packaging.

Particular thanks are due to Chris Ellison and the OPM Group for their valuable technical and market input and support, for reading, checking, correcting and amplifying the various book chapters, and assisting with illustrative material. Their input and encouragement has been an essential element in writing and finalizing this book for publication.

Chapter 1

———

Flexible packaging – an introduction

———

Flexible packaging is not new. It has been used for many different packaging applications, using a range of flexible packaging substrates and printing technologies, for well over 150 years and, today, is one of the fastest growing of all the printing and packaging processes. End uses can be found throughout the consumer retail products, industrial, agricultural, shipping, health and personal care, confectionery and frozen food sectors.

———

Perhaps best defined as packaging that is designed to hold products, goods, solids, liquids or powders and which is flexible in format, has no defined fixed shape and can be readily or easily changed, flexible packaging is usually supplied on rolls ready for an end-user or packer to form (into a tube, sachet or pouch, etc.), fill and seal (or close) at one or both ends, or to wraparound products.

Materials used for flexible packaging may be paper or paper-based materials, metallic foils, plastics films, regenerated cellulose film, or composite structures incorporating multi-layers, coatings, metalization, lamination and impregnation that provide specific barrier properties to gases, moisture, fats, light, oders, etc. New flexible packaging materials and constructions continue to evolve.

The first processing stage in the manufacture and use of flexible packaging (after origination and pre-press) will usually be printing on a web-fed reel-to-reel flexo or gravure press – mainly in multiple colors. There is also some use of web offset litho and screen process and, in recent years, the use of narrower web presses and digital printing for shorter runs, versions and variations. Around 60 percent of flexible packaging is printed direct onto the substrate surface; the other 40 percent is printed on the reverse of the substrate and laminated.

Apart from the printing process itself, flexible packaging substrates may be surface coated or treated to provide a moisture or vapor barrier, withstand lamination, sterilization, pasteurization or irradiation, provide strength, or make the pack sealable. They can also be shaped to match function, brand and appeal for a wide variety of packaging applications and markets.

A wide range of filling equipment for the flexible packaging market has been developed over the years, from simple manual machines for small packaging operations up to high-speed, fully-automated form, fill and seal lines for large volume packaging.

Rapid growth in flexible packaging in recent years has also been driven by a wide array of innovative new origination, print image carrier and conventional wide and narrow-web analogue and digital printing press technologies, as well as developments in flexible packaging substrates which have seen quality decorated films become ever more attractive in a consumer-driven world.

So where did flexible packaging have its early origins?

EVOLUTION OF FLEXIBLE PACKAGING

Flexible packaging using multiwall and laminated paper bag and sack constructions has been used for many decades, while the use of paper for the manufacture of flour and sugar bags can be originally traced back even further. Indeed salt, flour, rice and sugar bags have been in use for well over one hundred years, with the first patent for a machine to manufacture single-wall paper bags being granted way back in 1852.

Barrier paper constructions using multiple plies for strength and vapor-barrier performance, the addition of foil and plastic layers for moisture, gas, flavor and oder protection, and heat-sealing polyethylene layers were already coming into use by the mid-20th century. Even today, paper-based flexible packaging is still widely used and has attained growing interest as being more sustainable than some of the newer flexible packaging materials.

Outside of paper-based flexible packaging, the use of aluminum foil for flexible packaging developed rapidly after World War II when large quantities of aluminum foil became available for commercial use. Being compatible with foods and health products – offering superior flavor retention and a longer shelf life than other available flexible packaging materials – the use of aluminum foil soon found new applications and markets.

Flexible foil lids used to close and seal yogurt and cream pots and cheese dips were introduced in the United States in the mid-1960s. These provided a more reliable seal, improved shelf life, greater protection and tamper-evidence. Probably the only other material found in regular use for flexible packaging prior to the middle of the 20th century was regenerated cellulose film (RCF).

However, from the 1950s and 1960s, plastics started to play an increasing role in the world of packaging and, particularly, in the whole area of flexible packaging. It was from this time that (thin) plastics, biaxially-orientated films, extrusion coatings, film lamination, polymer adhesives and heat-seal coatings all evolved, along with new developments in converting equipment, flexographic and gravure printing machinery, ancillary materials, electronic pre-press and, more recently, digital printing, that has seen flexible packaging become one of the world's largest packaging types.

Today, depending on country or region, some 70 percent, or so, of overall flexible packaging industry revenue comes from polymer materials, especially Polyethylene (PE), Low Density Polyethylene (LDPE), High Density Polyethylene (HDPE), Biaxially-Orientated Polyethylene (BOPP) and Biaxially-Oriented Polyethylene terephthalate (BOPET). The remaining materials used for flexible packaging are paper or paper-based substrates, and aluminum foils and foil laminates.

THE FLEXIBLE PACKAGING MARKET

With more than a dozen market research and consultancy companies regularly undertaking studies of the global, regional or national markets for flexible packaging, there is certainly no shortage of data available. Unfortunately, there seems to be quite large variances between these companies on what is included as flexible packaging, in the value of the market, the volume of materials used and the forecast growth of the market up to 2022 or 2024. All of which prove somewhat inconclusive.

Growth forecasts set out by the majority of the industry studies currently available range from as low as +4.3 percent per annum up to as high as +5.2 percent per annum. For the purposes of this particular book a consensus average growth of 4.9 percent per annum has been used. Highest growth forecasts for flexible packaging are for the Asia Pacific region at +6 percent per annum (India on its own at up to +10 percent per annum), with Europe and North America at around 3.2 to 3.5 percent per annum.

To try and achieve some level of consensus on the current and future size of the global flexible packaging market the Label Academy has analyzed the majority of the currently available studies to produce its own market analysis based on an average overall figure. The results can be seen in the chart shown in Figure 1.1. As can be seen, the value of the global flexible packaging market by 2024 (based on the average of all the studies) is forecast to be in excess of 225.50 billion USD.

However, if only based on the highest market

VALUE OF GLOBAL FLEXIBLE PACKAGING MARKET 2015 TO 2024 BY VALUE IN USD BILLION

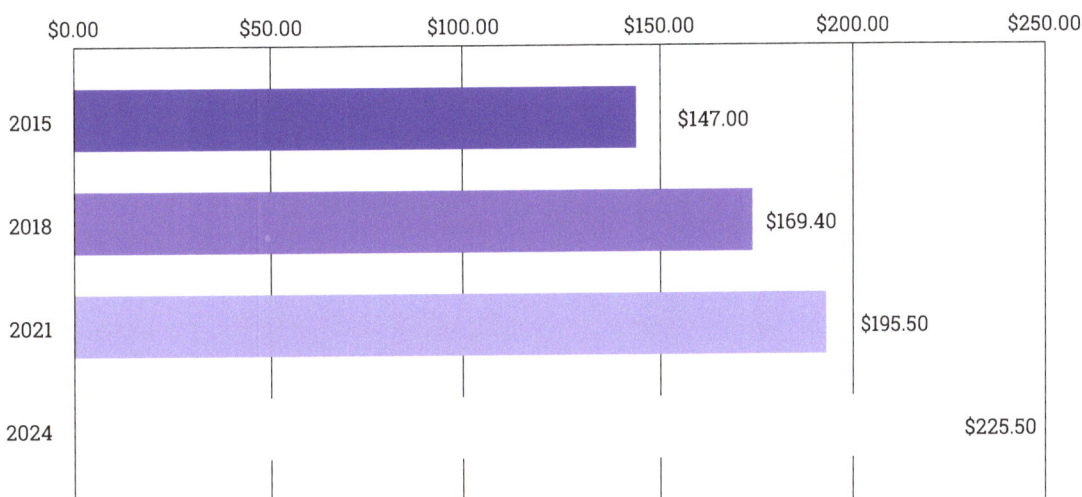

Figure 1.1 Global flexible packaging market to 2024 by value in USD billion. Source: Average of all major market studies

forecasts presented by one or two research companies, the flexible packaging market by 2024 could be as high as 300 billion USD, or more.

Many factors have an impact on the potential growth of flexible packaging – consumers looking for easy, lightweight, convenient and efficient packaging, more re-sealable packs, more user-friendly features, demand for convenience products, more single portion packs, and more product varieties, are just some of key influences driving growth in the food and beverage, healthcare, cosmetics, FCMG, consumer goods, confectionery, frozen foods and personal care sectors.

Stand-up pouches, pillow pouches, disposable sachets and single portion packs are all seen as important growth segments for flexible packaging, as is the growing use of flexible packaging in the healthcare sector. Rising packaged food demand and a growing preference for flexible packaging over rigid packaging materials are all seen as key factors.

Standing against all the growth factors are increasing pressures on plastic packaging waste, demand for more sustainable packaged products,

more and easier recycling, scarcer resources and the rising price of oil, petroleum and polymer products.

In general, the flexible packaging market can be segmented by the type of material used, by the type of product produced, by the printing technology and/or converting process used and by end-use applications. Each of these segments is introduced in this chapter and will then be examined in more detail in the subsequent chapters of the book. So let's commence with an introductory guide to the range and variety of flexible packaging materials and constructions.

FLEXIBLE PACKAGING MATERIALS USAGE
As mentioned earlier, flexible packaging makes use of quite a wide range of materials that include paper, metallic foil, regenerated cellulose film and plastics. Single and multi-layer papers and boards were the earliest materials used, followed by metal foils and regenerated cellulose film and, most recently, plastics films. Additional properties can be achieved by combining materials, such as an aluminum layer on a

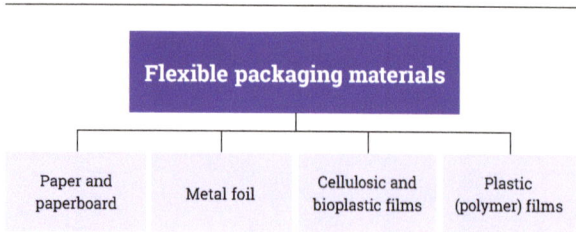

Figure 1.2 The main types of materials used for flexible packaging

polymer surface, which then offers new properties of oxygen and water transmission resistance. However, the basic segmentation of flexible packaging materials, can be quite simply shown in the flow chart in Figure 1.2.

With different products requiring different types of protection it means that some flexible packaging is made from a single layer material while, in other cases, multi-layer materials are required to provide the appropriate barrier and protection. In multiple-materials packaging, each layer performs a different function in protecting and preserving the product. By using materials with properties geared toward specific performance, manufacturers can meet customers' varying needs including product protection, contamination prevention, extended freshness, puncture, tear and burst resistance, tensile strength, and seal strength

In addition to these multi-material constructions, coatings and surface treatments have been developed to provide special barrier or resistance properties, add low temperature performance, increase durability and provide sealability. The emergence of biodegradable and bioplastic films has also been a key factor in the growth of flexible packaging.

As narrow-web printers move into flexible packaging they must also understand the requirements of sealant layers and make sure they understand how the film will be sealed, and which side seals to which side, so that they can purchase the appropriate films. Most flexible packaging films are heat sealed, but not all sealant layers are compatible. Generally sealant layers will always be compatible with themselves.

Substantial barrier properties can be achieved through coatings or metalization. Film manufacturers

have been applying these for years. However, converters today do have the opportunity to apply their own barrier to film. Newer types of coatings are available which allows gravure and standard flexo converters to achieve the highest levels of barrier in a flexible coating that is not subject to flex cracking. The only word of caution is that the converter then becomes responsible for the barrier and the protection of the product in the package, and there is a risk of very high claim levels if the film underperforms.

While there are a range of flexible packaging materials and constructions that are available and widely used, it is plastics (polymer) films which now dominate the market and are said to currently account for over 70 percent of the industry's revenue. Commonly used plastic films include Polyethylene, Polypropylene, Poly Vinyl Chloride, Polyamide and Ethylene Vinyl Alcohol. Again, this range of basic filmic flexible packaging materials – and some of their variations – can be shown in a simplified flow chart format, as shown in Figure 1.3.

The choice of flexible packaging material, the number of layers to be used, the specific laminate construction, the use of barrier or protective coatings, the need for a sealer, etc., will be determined by the specific pack properties that are required and the demands that will be placed on the material when it used for specific goods or applications. Light, moisture and oxygen can hugely effect a product's flavor and freshness, so protecting it from these elements is also important.

In general, the selection of a flexible packaging material is most likely to be related to what it is to be used for. This is indicated in a simplified form in the examples shown in Figure 1.4.

The types of films and their usage will be amplified in more detail in Chapter 2.

TYPES OF FLEXIBLE PACKAGING

The manufacturers of all kinds of products and goods – whether consumer or industrial – expend a vast amount of time and effort in researching, designing, manufacturing and packaging their goods, so will be looking for the best packaging solution to take those goods to market, to provide protection in transit and in store, to keep the products fresh and long-lasting, and to display them in the best possible way with optimum shelf impact.

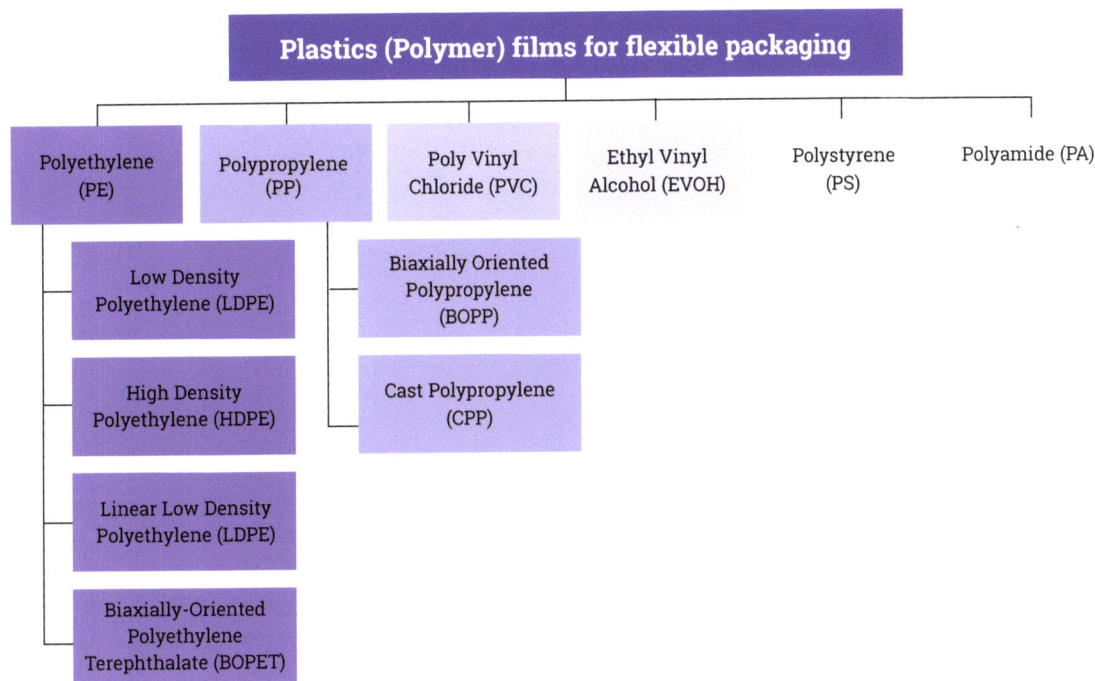

Figure 1.3 Plastics films used for flexible packaging

Many products (especially foods) require protection that can be provided by flexible packaging incorporating barrier layers and/or coatings. Although not exhaustive, the following list provides a general guide to why different types of flexible packing materials and layers are necessary:

- **Bakery** – Degraded by loss of humidity, however water retention can also cause a loss of crispiness, so perforated films are used for crusty products.
- **Biscuits** – Generally degraded by humidity uptake, leading to loss of crispiness. Complex products with chocolate and cream are degraded by oxidation, odor loss or uptake.
- **Chocolate confectionery** – Can be degraded by: moisture/humidity, which causes sugar bloom; odor, often coming from the inks or the use of recycled board; insects, if poorly sealed; and light/oxygen, which causes rancidity.
- **Dehydrated food and beverage** – Generally, these products have a very long shelf life and

Usage of flexible packaging materials

Plastic films
Used for wrapping fresh produce, dry foods, meat and cheese, freezer packs, carrier bags, or foods on high-speed packaging lines

Aluminum foils
Used for confectionery, ready meals, pharmaceuticals, soups and sauces, preserved foods and liquid foods

Paper and paperboard materials
Used for a wide range of paper bags and sacks

Cellophane film
Chocolate, toys, gifts, pharmaceuticals and cigars

Figure 1.4 A guide to flexible packaging materials and their key uses

Types of products flexible packaging may contain				
Solids in the form of powders, granules, grains etc.	Solids in the form of blocks, bars, slices, etc.	Individual items or groups of items	Liquids, creams, pastes and gels	Medical devices, kits and dressings, etc.

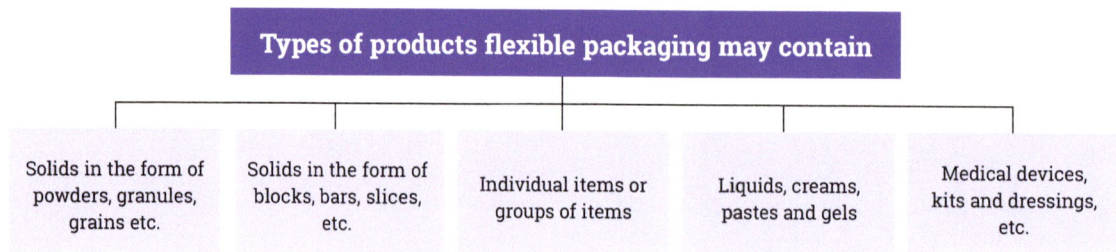

Figure 1.5 Some of the many types of products that flexible packaging may be asked to contain

therefore require a very high barrier to water, aroma and oxygen (when gas flushed).
- **Pet food** – When wet, pet food is degraded by oxygen, light and loss of aroma and/or contamination by odor.
- **Sugar confectionery** – When uncrystallized, products tend to absorb water. When crystallized, they tend to lose moisture.
- **Chips and snacks** – Degrade through rancidity, which necessitates a barrier to oxygen and light, and loss of crispiness, which is solved by providing a moisture barrier.

It can perhaps be seen from the list why flexible packaging has been, and still is, one of the fastest growing of the packaging technologies. It's a simple, hygienic and cost-effective way of protecting and closing or sealing many different types, shapes and products – whether solids, powders, liquids, creams, gels, pastes or shapes. Quite simply, what flexible packaging can be used to contain is shown in Figure 1.5.

With such a wide range of products and goods being packaged, the range of flexible packaging types and solutions is of necessity extremely diverse and will include:
- Roll-fed film
- Wrappers
- Bags
- Sacks
- Packets
- Sachets
- Pouches
- Lidding

- Multi-layered structures
- Zippered, pourable and re-closable packs.

Examples of some common flexible packaging types are shown in Figure 1.6.

There is little doubt that providing added-value features can significantly enhance the value to a customer's flexible packaging. Such features can be re-closure capabilities, flip-top caps, pumps, easy-carry handle options, gussets, squeezability, hang hole packs, easy-open features, notch options, dispensing possibilities, premade straw holes, self-heating pouches, aseptic pouches, child-resistance, tamper-evidence, pourable spouts, zippers, and tear-off pull tabs.

Depending on the particular reference source, flexible packaging may also be shown as including both shrink and stretch wrap, shrink sleeve and stretch sleeve applications. The largest market for all the above different types of flexible packaging solutions can be found in food (both retail and institutional), which accounts for about 60 percent of all shipments.

New developments in flexible packaging include the evolution of spouted packages. Spouted pouches are re-closable, which makes them a good match for grab-and-go products. The pouches can even be made with die-cut handles for better portability. In addition to being customizable, re-closable, and portable, spouted pouches offer great flexibility in the types of products they can hold, including food, beverage, condiments, dry mixes, pet food, granulates, and powders.

Pillow bag

Gusseted bag

Stand-up pouch

Tetrahedral bag

Quattro seal bag

Quattro seal bag with back seal

Sachet

Diaper bag

Ponytail bag

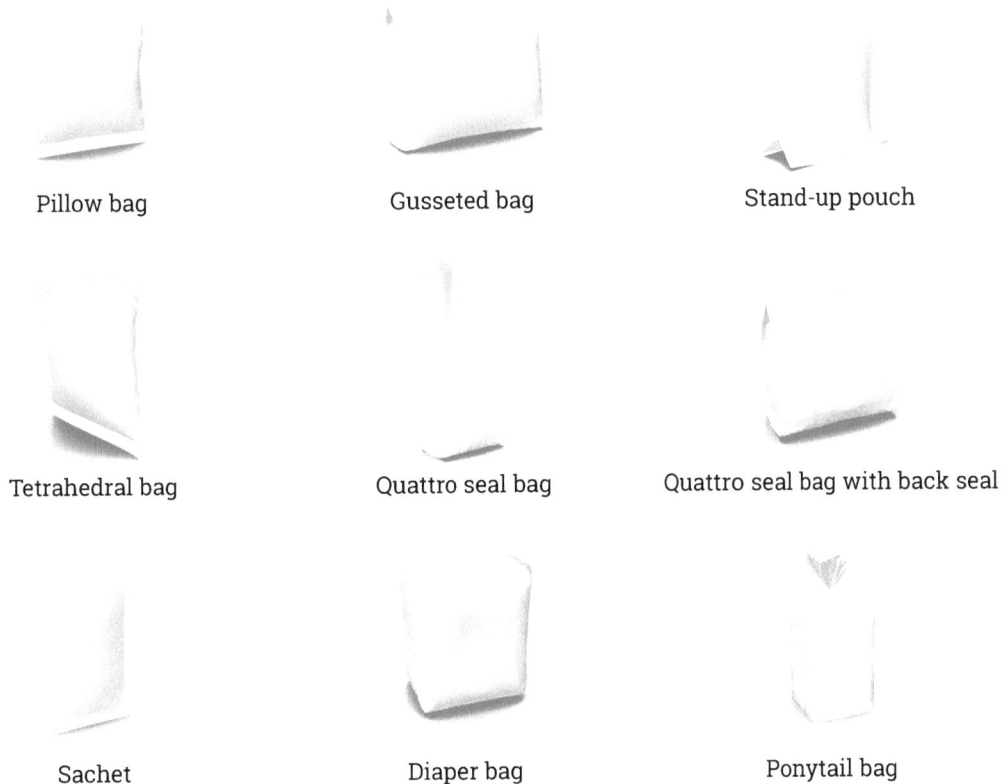

Figure 1.6 Image examples of common flexible packaging types. Source: Esko

Worldwide, many hundreds of examples of innovation in flexible packaging can be found. Each one starts from an idea; meat should stay fresher longer, shipping costs should be lower, guarantee no noxious substances will seep into the product, and medicines should be safer for the consumer. Creativity and innovation have long been at the forefront of the industry.

Chapter 3 of this book will examine and describe in more detail some of the main different types of flexible packaging products and solutions..

PRINTING AND CONVERTING OF FLEXIBLE PACKAGING

Printing and converting of flexible packaging is predominately undertaken on wide-web (reel-fed) presses, mainly in multiple colors. Flexo and gravure are the dominant printing processes, but there is also some use of web-offset and screen printing. Historically, gravure has been regarded as producing the highest print quality. Combinations of printing processes (conventional and digital) in one press line are also becoming more common.

In more recent years, with shorter print runs, more versions and variations, smaller sizes and personalization, there has been growing usage of narrow- and mid-web flexo printing and, over the past few years, increasing interest and application of digital printing, with particular growing interest in UV inkjet printing – a process in which the UV inkjet inks are chemically very similar to UV flexo inks and therefore able to print on virtually the same range of materials

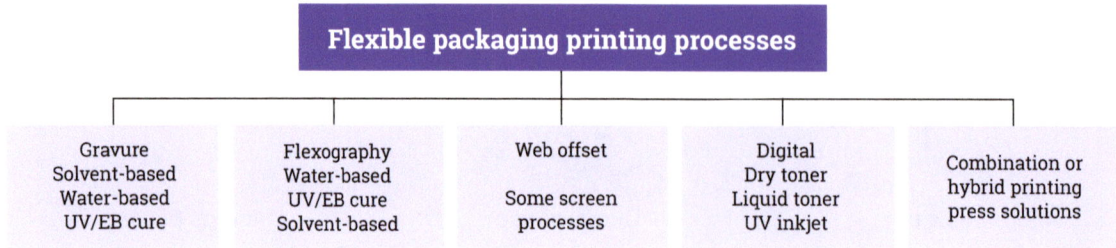

Figure 1.7 Printing processes used for flexible packaging

as UV flexo. The key flexible packaging printing processes are shown in Figure 1.7.

Traditional flexible packaging run lengths continue to trend downwards as wide-web converters struggle to print the smaller run lengths economically. There are however many narrower web in-line style printing presses (servo-driven machines in narrow-web widths) and new automated control systems that are now able to handle the shorter runs with a high degree of flexibility in terms of color sequence, configuration and available printing technologies. Servo drives and controls have also helped overcome the main challenges that have historically plagued the wide-web flexo market, such as gear marking.

Narrow-web flexo pre-press cost and quality can undoubtedly now successfully compete with more expensive wide-web solutions, and offer the possibility to combine printing of pressure-sensitive labels with unsupported film. Combination presses (flexo, offset, gravure), plus the possibility to add finishing processes like in-line lamination or cold foil raise the spectrum of printable products

As already mentioned, some 60 percent of flexible packaging is printed direct onto the (film) surface, with the remaining 40 percent printed on the reverse and laminated. Where narrow-web and digital printing is undertaken this is most usually undertaken on mono-layer flexible packaging materials, rather than on wider format complex multi-layer, laminating and sophisticated converting lines designed to provide a moisture and vapor barrier or withstand sterilization, pasteurisation and thermo-sealing.

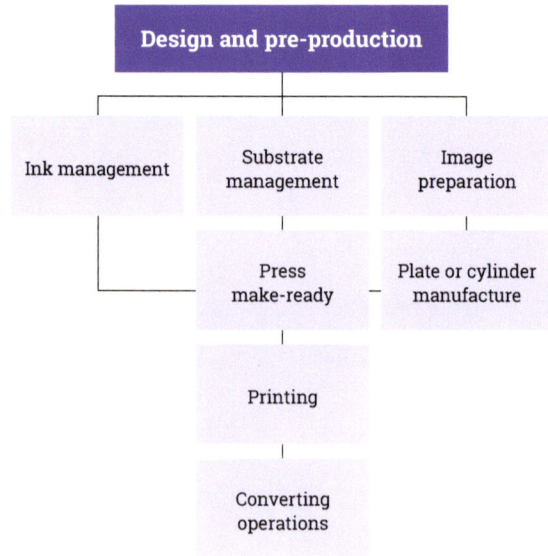

Figure 1.8 The design to production stages in the conventional printing of flexible packaging

When printing by any one of the conventional analogue printing processes the main stages in the whole design to print operation can be seen in Figure 1.8.

The printing of films, some often quite thin, means that presses may require sophisticated tension control, perhaps heat management systems, static control, corona treating and registration, etc.

Print and packaging – flexible packaging

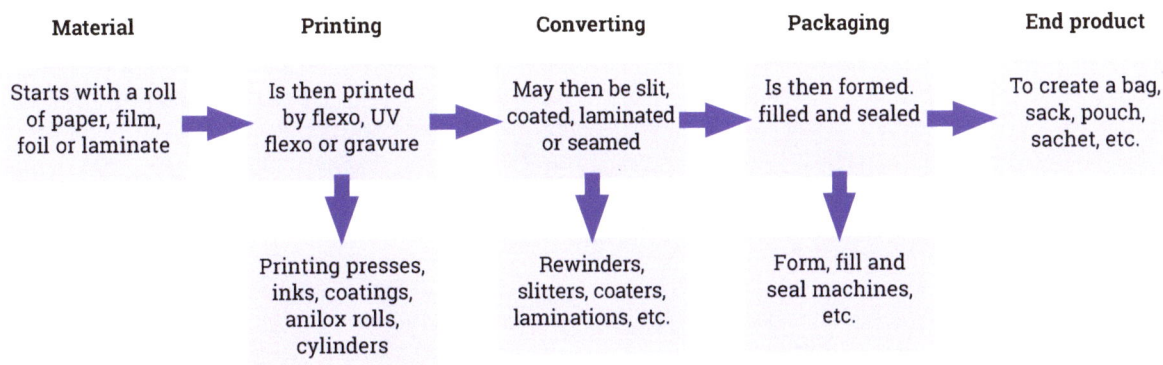

Material	Printing	Converting	Packaging	End product
Starts with a roll of paper, film, foil or laminate	Is then printed by flexo, UV flexo or gravure	May then be slit, coated, laminated or seamed	Is then formed, filled and sealed	To create a bag, sack, pouch, sachet, etc.
	Printing presses, inks, coatings, anilox rolls, cylinders	Rewinders, slitters, coaters, laminations, etc.	Form, fill and seal machines, etc.	

Figure 1.9 Typical stages in the converting of flexible packaging

Food Labeling – Food labeling regulations

- Traceability measures
- Food contact
- Materials used
- Safety/usage Instructions
- Information not misleading
- Ingredients
- Languages required

Figure 1.10 A guide to some of the most important aspects of food labeling regulations and directives

Typical converting operations found in flexible packaging production include varnishing or lacquering, surface treating, heat-seal or adhesive coating, priming, laminating and sealing. Coatings are often used to enhance the aesthetic appeal of the substrate, to protect the substrate and printed image, or to improve slip characteristics. Typical stages in the converting of flexible packaging are shown in Figure 1.9.

A barrier coating or sealing coat can be applied to prevent migration of ink, adhesive, or other substances through the face material. With flexible packaging, an anti-fog coating may be required to provide a specific barrier protection, such as moisture resistance or protection against light.

Laminating is the process through which two or more flexible packaging webs are joined together using a bonding agent to improve the appearance and barrier properties of the substrate.

Inks and varnishes obviously play a key role in not only providing the decorative and information performance of flexible packing, but also in adding the required barrier and sealing properties, coatings and seals. As most flexible packaging is used for food or for sensitive applications (body-care, cigarettes, etc.) it is important that neither the materials nor the inks should be able to contaminate the product by odor, migration or trace particles. Testing and test procedures are therefore important. See also Chapter 5.

Most countries have quite complex standards and regulations which determine which label and packaging materials may come into direct contact with food-stuffs and human skin, as well as a whole host of other food labeling requirements (see Figure 1.10). These products are prohibited along with those where there may be some transfer or migration of substances to food. The label user and/or label manufacturer are held liable for any failure on their part to comply with standards.

Before leaving this section, mention should perhaps be made regarding pouches. The majority of pouch manufacturing has historically been undertaken with form-fill seal machinery, which provides simultaneous pouch production, filling, and sealing on one piece of equipment. However, as pouches have become more complex, some packaged goods companies are turning to preformed pouches that are purchased direct from the printer/converter. Packaging machinery makers are adding features that can make two-part pouches, laminate multiple layers of films, and handle technologically advanced substrates.

When designing and manufacturing flexible packaging for sachets and pouches for liquids, testing is a key in preventing sachet or pouch failures. Throughout the stages of package engineering, packaging film manufacturing, and sachet/pouch printing and converting, the supplier should subject the pouches and packs to various types of testing, which may include:

- Product/package compatibility testing
- Compression testing
- Seal testing
- Puncture resistance testing
- Water bath testing
- Tensile testing
- Burst testing
- Drop testing.

Each type of testing plays a critical role in ensuring that the final package will function optimally. For example, when forming pouches, frequently performing water bath testing (a test using external pressure) and internal pressure testing throughout each pouch converting run can immediately detect any leaks or weak seals.

Chapter 4 of this book will discuss the pre-press, printing, converting and testing requirements for successful flexible packaging.

WHAT IS FLEXIBLE PACKAGING USED FOR?

With the introduction of ever-more filmic materials since the 1960s, as well as advances in coatings, surface treatments and seals, sachets and pouches, the range of applications and markets for flexible packaging has grown dramatically and now includes packaging for many different and varied markets. An indication of just how diverse flexible packaging has now become can be seen in the following, non-exhaustive, list of market applications:

- **Food**
  - Dairy and cheese produce
  - Frozen foods and ready meals
  - Meat, fish and poultry
  - Dehydrated and dry food – soup and sauce packets, rice, food mixes
  - Coffee and tea
  - Chips (crisps), snack and nuts
  - Ice cream novelties, ice lollies
  - Fresh fruits and vegetables
  - Spice foods
  - Bread and bakery products – cookies (biscuits) and cakes, baking ingredients
  - Chocolates, sweets
- **Personal care**
  - Toiletries and hygiene, shampoo, liquid soaps, creams, lotions, gels
  - Cosmetics
  - Beauty products, wipes, mud packs
  - Pharmaceutical, nutraceutical, veterinary, medical
  - Pharmaceutical
  - Nutritional
  - Animal care and veterinary products
  - Medical devices
- **Home and garden**
  - Lawn care and grass seed
  - Bark
  - Fertilizer
  - Laundry detergents
  - Paints, pastes, plaster
  - Household items
- **Animal care**
  - Pet foods and nutrition options
  - Pet care products
- **Beverages**
- **Automotive**
- **Agriculture**

Figure 1.11 Examples of printed filmic packaging used in the food sector

What is flexible packaging used for?

Types	Applications
Frozen food packs in supermarkets	Frozen vegetables, meat, fish, poultry, ready meals, etc.
Candy (sweet) wrappers	Mars, Twix, chocolate, etc., wrappers
Snack packs	Cookies (biscuits), chips (crisps), nuts, savouries
Garden center and horticultural packs	Fertilizer, bark, grass seed, lawn care
DIY packs	Wallpaper paste, plaster, etc.
Agricultural packs	Livestock feeds, pesticides, etc.
Stand-up pouches	Washing up liquids, laundry detergents, etc.
Dry food packs	Soup, coffee, sauces and food mixes
Bread wrappers	Bread, buns, cakes, etc.

Figure 1.12 Types of flexible packaging and their applications

As can be seen, the list covers almost all types of consumer food, pharmaceutical, health and beauty, nutritional, garden, DIY, leisure and other retail market applications, as well as the industrial, automotive, agricultural, horticultural and medical sectors.

This variety of packs, each designed for specific products and markets, typically take the shape of a bag, pouch, sachet, envelope, liner, or overwrap, and are defined as any package or any part of a package whose shape can be readily changed.

Leading the way in packaging innovation, flexible packaging adds value and marketability to food and non-food products alike. From ensuring food safety and extending shelf life, to providing even heating, barrier protection, ease of use, resealability and superb printability, the industry continues to advance at an unprecedented rate. Examples of printed filmic packaging used in the food sector can be seen in Figure 1.11.

Figure 1.12 aims to put some of the main end-usage markets into context in terms of the type of flexible packaging used, whether bags, sacks, wrappers, pouches, sachets, etc., and then relate these to an end-use application.

Chapter 7 will look in more detail at flexible packaging markets and applications and provide some case study examples.

WHAT ARE THE KEY ADVANTAGES OF FLEXIBLE PACKAGING?

As one of the fastest growing segments of the packaging industry, flexible packaging is able to combine the best qualities of plastic films, paper, cellulose and aluminum foil – plus varnishes, coatings, lacquers, adhesives, sealers, etc., – to deliver a very broad range of protective properties while employing a minimum of material.

The life cycle attributes of flexible packaging demonstrate many sustainable advantages. Flexible packaging starts with less material and less waste in the first place. Of all the packaging materials, flexible packaging has the smallest percentage weight – typically less than two percent – of the finished pack than other packaging materials (glass, cans, cardboard and corrugated), is easy to handle and greatly reduces landfill discards.

Innovation and technology have enabled flexible packaging manufacturers to use fewer natural resources in the creation of their packaging, and improvements in production processes have reduced water and energy consumption, greenhouse gas emissions and volatile organic compounds. Even more, lighter-weight flexible packaging results in less transportation-related energy and fossil fuel

Benefits and advantages of flexible packaging

- Flexible packaging materials are light in weight , easily transported and handled, easy to open, store and reseal
- They offer the lowest percentage of waste of almost all packaging materials
- Flexible packaging is widely converted into packaging for many diverse end-product categories – bags, sacks, sachets, pouches, wrappers, etc.
- It offers a variety of re-closure and dispensing options
- Flexible packaging requires less energy to manufacture and to transport than most other forms of packaging, and generates smaller quantities of greenhouse gases on its way to market
- High quality printing creates excellent shelf appeal and enables visibility of pack contents
- Flexible packaging offers consumer convenience
- Food packs indicate and maintain freshness, extending the shelf life of many products, especially food
- Flexible packaging provides an efficient product to pack ratio, uses less energy and creates fewer emissions

Figure 1.13 Benefits and advantages of flexible packaging

consumption, and environmental pollution.

With its flexibility and versatility, custom qualities, efficiency in conserving resources, and sustainability, it's not surprising that flexible packaging has experienced rapid growth in recent years, and is forecast to continue good – above GDP – growth levels for at least the next five, or more, years.

Flexible packaging is undoubtedly at the forefront of important packaging trends in product protection, packaging design and performance, consumer convenience, and sustainability which positively impacts the environment, consumers and businesses.

To summarize the benefits and advantages enjoyed by users of flexible packaging this chapter ends with a tabular review (Figure 1.13) of what has been presented so far during Chapter 1.

Quite simply, flexible packaging makes thousands of consumer and industrial products more convenient, enjoyable, and safer to use, enhanced by the commitment to innovation, technology, and sustainability that are the hallmark of flexible packaging suppliers.

Chapter 2

———

Paper and board, metallic foil, films and multi-layer constructions

———

With flexible packaging materials being used to wrap, or contain, so many different kinds of products – powders, liquids, gels, creams, solids, etc. – it is not surprizing that such a wide variety of papers, films, metallic, metalized and other substrates, as well as surface treatments and multi-layer and barrier constructions are required.

———

Depending on the particular products that require packaging, the materials used may require protection from the product inside the packs, such as moisture, oils or greases, chemicals, detergents, etc., or to protect the product from external influences, such as atmospheric impurities, light, UV rays, gas, oxygen, moisture again, puncturing or tearing. A product's formulation can undoubtedly be impacted adversely if any of these elements are able to migrate into or out of the package.

When looking at food packaging materials for example, the particular type of food product, its chemical make-up, size, expected storage conditions and required shelf life, moisture content, aroma/flavor and product appearance are all just a few of the many characteristics that must be taken into account when selecting the right material for a specific food or other product.

An ongoing trend in food packaging for example, is the way the pack is designed so as to extend the shelf life of foods while at the same time maintaining a fresh-like quality. Such requirements places a high demand on the selection of materials that not only need to provide the necessary properties to maintain the quality of the food but must also be done at an economic and cost-effective price.

Good design is of course critical in persuading consumers to purchase, but the repeat purchase will only be made when the promise of the design is matched by the quality of the product inside the packaging. The quality of the product inside the pack can only be guaranteed over the period of the shelf life by selecting a good-quality packaging paper, film, foil or barrier material and matching the capability of the material used to the protection requirements of the product. Get that step right, and the product will reach the consumer in the condition intended by the manufacturer.

Another aspect to be considered in food packaging is the permeability of the packaging material. This is one of the most critical features of the package that can impact on the quality of the food product. Materials can definitely be selected to provide a very long shelf life, but do they provide the best barrier. Added to that is the question of whether the extension of the shelf life justifies the cost of the material and the quality of the food? Knowing the key factors for material selection just based on permeability can be an essential part of the package design process.

Outside of foodstuffs, papers used for wrapping soaps and detergents for example, will need to protect the consumer from coming into contact with the chemical ingredients of these products. Hot melt coatings, anti-mold treatment and anti-fungus papers are all likely requirements for soaps and detergents. The papers will also have to perform well on high-speed packaging lines.

Enhanced storage and distribution requirements, retail and consumer shelf life demands, and the running of printed materials on filling, forming and sealing machines may also be important considerations when making the selection of which flexible packaging materials are to be used.

With so many varied and different requirements placed upon flexible packaging materials today, much innovation and development has been placed on the evolution of laminated films and high barrier constructions that prevent the permeation of water, water vapor, oil, oxygen, aroma, flavor, gas or light.

Lamination of a wide range of specific substrate types means that it is today possible to provide the best available packaging formats. Laminates now available provide the necessary strength and barrier properties, as well as print clarity, with encapsulated and laminated print eliminating scuff and offering peelable webs for consumer convenience. Each specific barrier material and formulation has its own unique barrier properties.

With printed flexible packaging, the mono-web, multi-web or specific barrier construction materials selected must also be able to provide printed shelf appeal, provide optimum printability, meet the necessary food or other labeling requirements, and perform as required in materials, print and usage tests.

In general, brand owners should know the performance and barrier properties required to protect their products. They have food scientists working for them, so any converter intending to make a speculative approach, or propose a new structure, will need to do their research carefully, and have an idea of what they are proposing and why it will work.

Look at the structures already being used in the market place; generally, they are used for good reason. If the product is successful it means that the packaging is probably fulfilling its purpose and preserving the product, allowing it to be enjoyed with the texture, flavor and aroma with which it left the factory.

Knowledge around barrier protection is undoubtedly crucial for servicing the diverse flexible packaging market. Suppliers must be able to provide the right sealant for a given application and be able to support clients with 'fitness for use' testing.

To summarize, what general information does the printer/converter need to understand before ordering flexible packaging materials? Shelf life, as already explained, is important and defined as the period in which consumer acceptability is maintained. The quality of most products changes over a period of time through changes to color, texture and flavor.

Shelf life is shortened by a number of factors, including: moisture, where a gain or loss can affect the texture and make a product go stale or go soft, or it can act as a catalyst to degradation in products containing fat; oxygen (which causes oxidation of products that contain fat or oil), and can assist in color changes and the onset of mold.

Light is another catalyst for oxidation that causes rancidity, and oils and fats to break down, causing odors and unwanted contaminating smells. Highly flavored foods are likely to lose aroma compared to bland foods that are likely to absorb odors. Additionally, the environment must also be considered. Secondary packaging, warehousing, transportation, distribution, temperature and in-store situations can all affect the shelf life of a product.

The table shown in Figure 2.1 is intended as a general guide to the list of food types that require protection, and shows why selection of the optimum flexible packaging material is important. It should be noted that products within these categories can vary in terms of need.

So, having discussed the basic requirements of flexible packaging and having a general knowledge of products that require protection, it now becomes possible to go back to the flow chart initially shown in Chapter 1 and now designated as Figure 2.2. This provides a basic guide to the main types of materials available – paper and paper-based materials, metallic foils, cellulose and plastics/polymer films.

Food and other products requiring protection

Bakery – Degraded by loss of humidity. However, water retention can also cause a loss of crispiness, so perforated films are used for crusty products

Biscuits – Generally degraded by humidity uptake, leading to loss of crispiness. Complex products with chocolate and cream are degraded by oxidation, odor loss or uptake

Chocolate confectionery – Can be degraded by: moisture/humidity, which causes sugar bloom; odor, often coming from the inks or the use of recycled board; insects, if poorly sealed; and light/oxygen, which causes rancidity

Dehydrated food and beverages – Generally, these products have a very long shelf life and therefore require a very high barrier to water, aroma and oxygen (when gas flushed)

Pet food – When wet, pet food is degraded by oxygen, light and loss of aroma and/or contamination by odor

Sugar confectionery – When uncrystalized, products tend to absorb water. When crystallized, they tend to lose moisture

Chips and snacks – Degrade through rancidity, which necessitates a barrier to oxygen and light, and loss of crispiness, which is solved by providing a moisture barrier

Figure 2.1 Food and other products requiring protection offered by flexible packaging materials and constructions

Let's now look in rather more detail at these main flexible packaging materials available to the converter and what they can provide, starting with the various paper and paper-based packaging materials. Please remember, a critical part of the flexible packaging process is selecting the optimum type of packaging material for the product, the printing process and the packaging line.

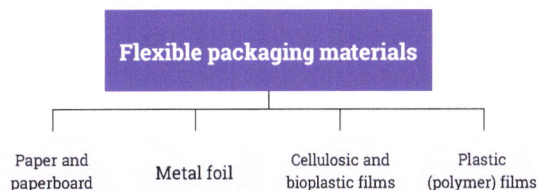

Figure 2.2 The main types of materials used for flexible packaging

FLEXIBLE PACKAGING PAPERS

While polymer films now dominate the usage of flexible packaging materials it has to be noted that papers continue to be important. They have a relatively low cost when compared with other materials. They are widely used in laminated webs (as light barrier materials), have a tactile feel and touch and offer good environmental performance, including low production energy levels, stiffness, breathability and cleanliness. They are additionally one of the easiest of the flexible packaging materials to recycle (over 60 percent re-use in Europe).

Various grades of paper are used for flexible packaging, representing different chemical (e.g., sizing) and physical (e.g. calendaring) treatments during paper making. Coatings on paper are used for visual and functional effects that enhance the material's utility in packages. Paper's dead fold and tear property combinations are unique in the array of substrates used in flexible packaging.

Paper properties in lamination with other materials – providing better stiffness and puncture resistance – can often provide cost effective and unique package performance attributes for customer applications. Along with Cellophane and some bioplastics, paper represents one of the few flexible packaging materials produced from renewable natural materials.

Flexible packaging paper grades can range from simple wrapping papers (frequently made from mixed recovered paper) to kraft papers, both of which are predominately supplied in rolls. They are made from various virgin pulps, from recycled fibers or from a mixture of chemical pulp and recycled fibers.

Whether coated or uncoated, matte or gloss, flexible packaging papers offer versatility alongside consistently high standards. Most grades are recyclable and available as FSC and PEFC certification on request, and there are grades suitable for a wide range of end-use applications – from shopping bags, food wrappers, dry and fatty food wrapping, small bags, soup packs, soaps, and tea envelopes to cigarette soft packs, inner liners and tobacco pouches.

Clay coated papers provide the ultimate printing surface and are used where multi-color printing is required for high-quality color printed results. They can also be laminated to a range of different films or foils to produce the finished packaging material.

In extrusion-coated formats, papers are used for the flexible packaging of dried food, prepared meals and savory snacks, while wrapping papers and paper sachets/pouches can be widely found in fast-food restaurants, cafés, coffee shops and for savoury snacks and baked products. Laminations with aluminum foil are also used within the food sector.

Food products that have perishable properties, such as butter, cheese, curd etc. are known as perishable food. Butter, soft spreads and cheeses are the best example of food items that all contain fats. When these products come into contact with moisture and atmosphere, they will lose their taste, color and aroma, requiring a quality food paper packaging material to keep all these intact. Uncoated grease and oil resistant wet-strength flexible packaging papers are also available for food applications.

On the other hand, food items like wheat flour, rice, tea, coffee, sugar, salt, seasoning, soup mix, snack foods, chips, noodles, etc. can be termed as non-perishable foods. These have a longer shelf life over other items such as curd or ghee but still they need quality packaging materials to further extend their shelf and consumer life term. Food paper grades must also fulfil the requirements of the relevant food packaging/labeling legislation.

Special paper packing materials can be produced and are available for most packaging purposes, e.g. greaseproof papers used for packing butter, margarine, meat, sausages, etc. Such papers include vegetable parchment and glassine and are provided with barrier polymers.

One-sided double coated flexible packaging papers, applicable for flexo, offset and gravure printing are available for use in both food and non-food applications, from dairy products and confectionery to tobacco. Such papers may have a rough back side, useful for lamination applications.

Vapor proof papers are papers that have been chemically treated or laminated with a vapor barrier that will resist the passage of gases or vapor through it, again, typically used for food packaging. Greaseproof wrapping papers are made from chemical wood pulps which are highly hydrated in order that the resulting paper may be resistant to oil and grease.

Outside of foodstuffs, papers used for wrapping soaps and detergents for example, will need to protect the consumer from coming into contact with the chemical ingredients of these products. Hot melt coatings, anti-mold treatment and anti-fungus papers are all likely requirements for soaps and detergents. The papers will also have to perform well on high-speed packaging lines.

When studying paper-based flexible packaging materials the key initial choice is usually between **kraft papers** – primarily used for paper bags and sacks – and **wrapping papers**, which are used on their own or in laminations with other materials, but particularly for food packaging applications.

Kraft paper sack and bag kraft are high strength papers made from sulfate pulp which are used to manufacture multi- and single-wall sacks and a range of bags, either in form-fill-seal applications or other automated packaging processes, or also in loose bags used to pack products at the point of sale, such as bakeries, street vendors, etc. Kraft bag and sack papers normally have a greater bulk, a high tensile strength, and a rougher surface than the more usual kraft wrapping papers which are used in a wide range of multi-substrate applications as well as in plain wrapping functions.

Pulp used to make kraft papers is stronger and darker than that made by other pulping processes, but it can be bleached to make a white pulp. Fully bleached kraft pulp is used to make high quality paper where strength, whiteness and resistance to yellowing are important. Sack kraft paper, is a porous kraft paper with high elasticity and high tear resistance, designed for packaging products with high demands for strength and durability. Figure 2.3 provides a summary of kraft paper properties, performance and applications.

FLEXIBLE PACKAGING MATERIALS – PAPER		
Material	Properties	Applications
Kraft papers	• High tensile strength • Good tear resistance • Good bulk • Rougher surface	• Single and multiwall paper bags and sacks • Point of sale bags, e.g. bakery products • Packaging with a high demand for strength and durability

Figure 2.3 Kraft paper properties, performance and applications

Candy wrapping paper and twisting paper are primarily thin 30–40 g/m2 kraft papers and are mostly flexo or offset printed. These papers require a good strength, with highly oriented fibers. Twisting paper is mostly opaque and often supercalendered.

CELLULOSE AND BIOPLASTICS FILMS

Prior to the 1950s there was no availability of large quantities of thermoplastic film. The major transparent packaging film used at that time was a non-thermoplastic regenerated cellulose film. Made by chemically regenerating a dissolved cellulosic compound into a thin film or sheet, cellulosic film was commercialized under the trade name of Cellophane.

While the development of petroleum-based plastic films began to erode cellulosic films from the 1960s onwards, they nevertheless still remain an important and viable flexible packaging material today. Indeed, in the light of the Paris Climate Agreement, which was aimed at strengthening the ability of countries to deal with the impacts of climate change, the ongoing development of environmentally sustainable packaging materials has led to the market for cellulosic, bio-based and PLA (Polylactic acid) materials now increasing slowly but steadily – largely due to ever-increasing environmental and legislative concerns, and due to the largely unpredictable price of petroleum oil.

Recent years have certainly seen extensive research and development undertaken into bio-based packaging materials, especially in Europe, although at the present time, despite the ready availability of several bio-based solutions, most packaging products still heavily rely on fossil-based (polymer) materials which continue to contribute to enhanced emissions of greenhouse gases and generate a lot of waste.

However, cellulosic flexible packaging materials are forecast to have a bright future in bio-based packaging applications. Cellulose is a highly hydrophilic material due to a great number of hydrogen groups on the surface. Made using a dissolving-grade sulphite wood pulp, steeped in caustic soda, shredded and then converted to film-ready viscose which, in turn, is pumped to the wet-end of a film forming machine, the resulting cellulosic film is wound on a core ready for coating, slitting, or both.

Softeners may be added for dimensional stability for when the film is used in unsupported form or to provide best durability for bag and pouch applications.

Cellulosic coatings and films efficiently prevent the permeation of oxygen, grease, and oils – so giving the film its broad packaging functionality – but, on the other hand, provide very poor moisture barrier properties as water easily breaks the hydrogen bonds that hold the chains together. At high humidity conditions, the cellulosic materials tend to swell as a result of moisture absorption.

Without adding moisture-proof coatings, the cellulosic film will tend to lose moisture and can become brittle and shrink. If there is excessive moisture pick-up the film will lose its, otherwise excellent, gas and aroma barrier properties.

If the cellulosic film is to be used on high-speed packaging machines then heat sealability in the coatings used will become an important requirement. Uncoated cellulosic film generally has limited packaging uses, such as decorative wrappings. To supply the best technical solution and functionality, high barrier films such as NatureFlex are laminated to an internal sealing bio-polymer so that the final structure, certified compostable, can be heat sealed to compostable base trays made from bio-polymers or wood pulp.

NatureFlex films are produced from sustainable and responsibly sourced wood pulp harvested from managed plantations and are certified to both EU (EN13432) and US (ASTM D6400) composting standards. They are considered as the next generation of Cellophane film. In addition to industrial composting, the product has reached the standard required for home composting. NatureFlex films are engineered to provide a high barrier to moisture, aroma and gasses, have excellent transparency and high gloss; making them an ideal solution for a number of flexible packaging applications, compostable lidding structures, and as barrier layers for biolaminates.

More recently, a new generation of biodegradable films made from renewable resources (PLA bio-based resin) has been introduced as a key step in driving the flexible packaging industry towards even more sustainable solutions. The films offer the possibility to choose natural products and contribute to the reduction of greenhouse gas emissions and post-consumer waste.

In many applications today, biaxially-oriented PLA films can replace oil-based plastics like polypropylene, polyester or polyethylene, so providing two key environmental advantages: bio-based origin, certified by Vinçotte, and compostability according to the EN13432 norm.

The range of PLA films includes heat sealable transparent and solid white films and are designed to cover a wide range of food and non-food packaging applications, using existing converting and packaging technologies. Applications for PLA films are expected to continue to increase steadily.

Bioplastics are rapidly becoming one of the key raw materials to be used by flexible packaging manufacturers. Growing demand for bio-based PLA films in food, bakery, confectionery and snack packaging applications, owing to their easy recyclability and biodegradable nature, is expected to be a key market driver over the next seven years.

Of the newer bioplastic film materials, Cellulose nanofibrils (CNF), also referred to as nanocellulose, are seen as one of the most promising candidates for use as a sustainable material in the packaging industry. As well as being completely renewable, biodegradable, and safe to use CNF also possess exceptional physical and chemical properties. Unfortunately, the strength and barrier properties are highly dependent on the humidity conditions, hence further surface modifications have been developed to strengthen the position of

FLEXIBLE PACKAGING MATERIALS – CELLULOSIC AND BIOPLASTIC FILMS		
Material	**Properties**	**Applications**
Cellulose, bio-based and PLA films	• Good at preventing oxygen, grease and oil permeation • Poor moisture barrier without coating or lamination • Biodegradable, compostable, recyclable and sustainable	• Growing use in the food, bakery, confectionery, and snack packaging sectors • Compostable lidding structures

Figure 2.4 Properties and applications for cellulosic, bio-based and PLA films

cellulosic films in high-performance film applications.

A summary of the properties and applications for cellulosic, bio-based and PLA films are shown in Figure 2.4.

ALUMINUM FOIL AND METALIZED SUBSTRATES

Aluminum foils are used where their better barrier properties and aesthetic appeal give them the edge over flexible films or papers and are widely used in multi-layer flexible packaging laminations for products such as confectionery, ready-meals, pharmaceuticals, soups and sauces, preserved foods and liquid foods.

Foil has the best deadfold properties of the flexible substrates and provides an excellent 100 percent barrier to all gases and to moisture. It is also a good reflector of radiant heat.

For guidance on the thicknesses of foils it should be noted that household foil is typically 17.5 μm (0.0005 inches), but it can be made available in gauges as low as 7 μm (0.00028 inches). Below a gauge of 12 μm (0.0005 inches) pin holing starts to

Flexible packaging material	Water vapor	Gas barrier	Odor Barrier	Water resistance	Oil Resistance
Aluminum foil	🟩	🟩	🟩	🟩	🟩
Polyethylene LDPE	🟦	🟨	🟨	🟩	🟥
Polyester	🟨	🟥	🟦	🟩	🟦
PVDC coated Paper	🟦	🟨	🟨	🟥	🟥
PE coated Paper	🟨	🟧	🟧	🟦	🟥
Waxed paper	🟦	🟦	🟦	🟦	🟦

🟩 Excellent   🟦 Good   🟥 Average   🟨 Poor   🟧 Very poor   🟥 Bad

Figure 2.5 Comparison guidelines of the transmission rates for various flexible packaging materials

become likely. Aluminum foil is also susceptible to flex cracking when folded. Consequently, most foils are supported with plastic and/or paper.

Consequently, foil may be readily combined into multi-layer structures with any of the other flexible packaging materials to form composites that are then able to meet a range of specific end-use requirements. For example, the oxygen barrier is significantly improved – up to fifty times for BOPP films, and up to ten times for BOPET films, which are two of the most commonly metalized flexible packaging films, along with BOPA (nylon) films. These three films, and other flexible packaging films, will be described later.

Aluminum foil is also used to seal the opening on plastic jars as well as acting as a very effective tamper-proof lid. Aluminum foil die-cut lids prevent the product from deterioration and pilferage during distribution and storage. They can be crimped over the top of the jar or, more frequently, heat sealed to provide a more secure closure and sealing system.

For instance, in Figure 2.5, aluminum foil scores the highest in all the parameters that determine the shelf life of products, which clearly suggests that any laminate structure involving aluminum foil as one of the layers will be best in terms of barrier properties and shelf life of the product.

The use of aluminum foil lids offers secure seals for sealing plastic container for valuable products packed in a variety of plastic jars and containers, thereby providing products with super protection against pilferage and making them tamper proof. Applications for aluminum foil lids are as diverse as food stuffs, cosmetics, beverages, flavored drinks, powders and yogurt.

The properties and applications for aluminum foils and metalized papers and films are summarized in Figure 2.6.

PLASTICS/POLYMERS FILMS
The primary reason for the increasing popularity and growth of plastics in flexible packaging is the highly versatile nature of a wide variety of polymer films now available, which enables them to be converted into a large number of shapes, sizes, and designs, with a whole portfolio of performance characteristics. Also, plastics are more flexible, durable, and cost-effective than many other materials used for flexible packaging which has led to their increased adoption.

Film flexible packaging options available to the converter include both single and multi-layer films: film/foil/paper laminates; metalized films, film sealing materials; form-fill seal processes and pouch styles. All these options will be discussed, either in this

FLEXIBLE PACKAGING MATERIALS – ALUMINUM FOILS AND METALIZED SUBSTRATES		
Material	**Properties**	**Applications**
Aluminum foils and metalized papers and films	• 100% barrier to all gases and to moisture • Good reflector of radiant heat • Readily combined with other substrates into high barrier performance multi-layer structures • Widely used as metalization on papers and films to provide a good light and performance enhancing barrier NOTE: Metalized films alone do not provide an acceptable aroma or odor barrier	• Confectionery, ready meals, pharmaceuticals, soups, sauces, preserved and liquid foods • Point of sale bags, e.g. bakery products • Tamper-proof lidding for yogurts, flavoured drinks, cosmetics, water pots, food stuffs, powders and curds

Figure 2.6 Properties and applications for aluminum foils and metalized papers and films

chapter or a subsequent chapter of the book.

A great many of the flexible packaging films used today are biaxially-oriented. Indeed, most plastic packaging films tend to be oriented. The orientation process simply consists of stretching the film in one (monoaxial, eg. OPP) or two (biaxial, e.g. BOPP) directions. Orientation dramatically improves film properties, such as stiffness and tensile strength, while at the same time reducing elongation. Orientation also increases film yield.

Biaxially-oriented films undoubtedly play a major role in the flexible packaging industry, due to the unique combination of mechanical, optical and barrier properties they are able to offer. Between them, these films provide the state of the art, as well as the ongoing future trends, for the most important of the flexible packaging film types, such as BOPP, BOPET, BOPA, BOPS, BOPE and BOPLA. Each of these film types will be expanded upon as the chapter progresses.

Of the biaxially-oriented films mentioned, BOPET and BOPP in particular, can be regarded as the workhorses of flexible packaging films, principally in laminations, where they can be used in the as-oriented form, but are often vacuum-metalized for applications where the films themselves do not offer sufficient barrier protection. BOPP is widely used in salty, dry snack packaging – as it offers better moisture vapor barrier than BOPET. BOPP also has the lowest density of the commonly oriented packaging films.

In addition, there are various additives that can be introduced into flexible packaging films during the film manufacturing process. These include:

• Process stabilizers (heat stabilizers)
• Environmental stabilizers (anti-oxidants, UV stabilizers)
• Surface modifiers (slip agents, anti-static, anti-blocking)
• Optical modifiers (pigments, nucleating agents)
• Functional additives (mechanical property enhancers).

This means that although a particular film type may be supplied by different manufacturers, they may not all have used the same additives in their manufacturing stages. Trials and testing of, apparently, the same films may be required to determine which performs the best in a particular application.

It should also be noted that polymer films perform differently at different maximum temperatures. This means that a low maximum temperature film will not perform well in an application where the packaged product film needs to perform at higher temperatures. For example, low density polyethylene (LDPE) has a maximum use temperature of 66 degrees Celsius, while polyethylene terephthalate (PET) provides a maximum use temperature of just over 200 degrees Celsius. High density polyethylene (HDPE) has a maximum use temperature of around 100 degrees Celsius.

To provide some useful guidance for the converter, particularly the narrow- and mid-web converter perhaps coming relatively new to flexible packaging,

FLEXIBLE PACKAGING MATERIALS – BIAXIALLY-ORIENTED POLYETHYLENE TEREPHTHALATE (BOPET) FILMS		
Material	Properties	Applications
BOPET film	• Superior stiffness • Chemically inert • Good heat resistance • Good oxygen, moisture and gas barrier properties • Wide operating window • High tensile strength	• Surface film of choice for retort pouches • Lidding material (white) used for dairy goods, such as yogurts • Lidding material (clear film) used for fresh or frozen ready meals • Used to protect food against oxidation and aroma loss when metalized in products such as coffee and pouches for convenience foods

Figure 2.7 Properties and applications for BOPET films

the following pages of this chapter set-out more on the nature, performance and uses for the key flexible packaging films types, together with mention of some of the more specialized film types.

BOPET (Biaxially-oriented polyethylene terephthalate) is a popular surface film in laminations where its superior stiffness, chemical inertness, heat-resistance, and oxygen-barrier properties – and a wide operating window (compared to BOPP) – make it a good choice for a wide variety of products. Produced by biaxial orientation of PET resin, BOPET has proved to be particularly suitable for converting and various other industrial applications.

Available as a clear, transparent or translucent material, BOPET film is manufactured commercially in a range of widths, thicknesses and properties depending upon the needs of end users. It can be

made as a single layer or can be co-extruded with other co-polymers into a multilayer film encompassing the desired characteristics of each material. They are also said to offer the highest tensile strength of all the packaging polymers. They have good moisture and gas barrier properties and low elongation.

Biaxial orientation of PET film makes it suitable for such applications as food packaging by increasing the product's crystallinity and thereby improving its tensile strength, heat resistance, and gas-barrier properties. The distinct physical properties of various types of PET film can be imparted into the product either during the polymerization of the PET resin, by the addition of chemicals such as slip modifiers (surface modifiers) or color additives, or subsequently during the PET film production process where various surface finishes may also be imparted by externally treating the film's surface(s).

More stable through printing and laminating processes than BOPP, it is often preferred where high-quality graphics are required. It is the surface film of choice for retort pouches because of its dimensional stability through retort

The improvements in mechanical, optical, and barrier properties of oriented films makes them compelling choices for flexible packaging structures. Of the biaxially-oriented films, BOPET offers performance at low thickness, high stiffness, good heat-resistance, and a reasonable balance of oxygen and moisture vapor barrier.

The largest application of thin BOPET films is in flexible packaging. White BOPET is used as a lidding material for dairy goods such as yogurts; clear BOPET finds applications in lidding for fresh or frozen ready meals; laminates containing metalized BOPET are used to protect food against oxidation and aroma loss in products such as coffee packaging and pouches for convenience foods. A summary of the properties and applications for BOPET films is shown in Figure 2.7.

BOPP (Biaxially-oriented polypropylene) is another of the commonly-used plastic film materials, this time made from stretched polypropylene (PP) resin. It offers superior moisture vapor barrier properties and has the lowest density of the BOPET, BOPP, and BON triumvirate. Against this, it has a poor oxygen barrier.

BOPP itself is a low density high performance film and has excellent mechanical, optical and barrier properties and can be regarded as one of the most

FLEXIBLE PACKAGING MATERIALS – BIAXIALLY-ORIENTED POLYPROPYLENE (BOPP) FILMS		
Material	**Properties**	**Applications**
BOPP films	• Superior moisture vapor properties • Excellent mechanical, optical and barrier properties • Easy processability • Good chemical compatibility • Poor oxygen barrier • Suitable for hot filling	• Flexible food packaging – dairy products, confectionery, etc. • Food pouches and gusseted bags • Clear wraps • Snack foods • Personal care and hygiene products • Pesticides • Often metalized for barrier performance

Figure 2.8 Properties and applications for BOPP films

efficient and competitive biaxially-oriented film solutions for many flexible packaging applications.

With a relatively low cost, easy processability and good chemical compatibility, BOPP films are an attractive product for food pouches and bags, clear wraps and most snack foods. With a higher softening point than, say, polyethylene (PE) it becomes suitable for hot filling. The film is often metalized and printed.

Recent studies predict that BOPP films are forecast to grow at 5.8 percent CAGR up to 2024, with the food packaging sector anticipated to be the fastest growing application segment, driven by the high demand for BOPP films for the production of a wide variety of package types and labels required in the food industry and tapes for industrial packaging purposes.

Variations of BOPP films include transparent, metalized, solid white, cavitated and matte films and

these are available with certificates of compliance with EU and USA legislation for material intended to come into contact with food.

A summary of the properties and applications for BOPP films is shown in Figure 2.8.

BOPE (Biaxially-oriented polyethylene). Before looking at BOPE films it is perhaps useful to have a better understanding of polyethylene itself. Polyethylene is a film material manufactured from ethylene – predominately from natural gas or petroleum – and, although it can be produced from renewable resources, it is not readily biodegradable.

Polyethylene packaging doesn't allow water vapor to pass through, meaning it can seal easily contaminated products away from dangerous elements. Many kinds of polyethylene packaging can be heat sealed, meaning the material can be wrapped tightly around the product and secured with an airtight seal.

Overall, polyethylene is the most commonly used plastic in the world and there are many variations and grades. Different grades of polyethylene can each have a different molecular arrangement that enables the production of different families of PE films – low-density polyethylene (LDPE), high-density polyethylene (HDPE) and linear low-density polyethylene (LLDPE).

Higher density polymers have their molecules more closely packed and are stiffer in nature, while low density polymers have loosely packed molecules are more flexible. Such are the properties and applications of LDPE, LLDPE and HDPE, that they are probably best considered as different films with different characteristics. Each of the films can be biaxially-oriented.

LDPE films are relatively low cost and easily processed. They are soft and clear, have the lowest softening and melting point (making them good for heat-sealing), are compatible with most foods and household chemicals, provide a fair moisture barrier (but poor oxygen barrier) and have a very high elongation. Being a very flexible material LDPE is particularly suitable for applications like shopping bags. Indeed, everything from food products to nails are packed in polyethylene bags – perhaps more commonly known as Poly Bags.

LDPE-based flexible packaging films are regarded as ideal for perishable goods, like packaging of food items, pharmaceuticals, liquids, pesticides, spices, edibles, industry goods, cosmetics etc. LDPE products are durable and moisture proof and are frequently combined with other films (such as PP) to give them heat sealability.

LLDPE film is very similar to LDPE, but with the added advantage that the properties of the film can be easily changed by adjusting the formula constituents. The overall production process is less energy intensive than LDPE. The film provides more stretchability and is tougher than LDPE. It also has a better heat-seal strength.

HDPE is a strong, high density, moderately stiff plastic with a higher temperature resistance and much improved water vapor barrier than LDPE, but is hazier. It is frequently used for applications such as laundry detergents.

Overall, polyethylene resins can provide a range of benefits, including: helping to keep food fresh, enhancing barrier properties and providing product protection. It's also inexpensive and highly chemically resistant. It can either be used by itself or combined with other substrates to provide barrier performance. High barrier structures can be produced by co-extruding PE, and PE with high barrier materials, such as nylon.

BOPA (Biaxially-oriented polyamide) is a widely used resin for the production of flexible packaging films, providing the best oxygen barrier performance, a good barrier to chemicals and aroma substances, and exceptional toughness and puncture-resistance. Standard versions of BOPA films, which tend to offer a good balance in mechanical properties between MD and CD, are widely sourced and specified for their unique contributions to packages.

BOPA films are increasingly in demand and widely used or the packaging of perishable food (especially fatty and oily food) due to its unique combination of properties. It is a highly transparent packaging film, which is also used for distilled goods packaging, agricultural products packaging or medical products.

A commonly used generic name for the polyamide family of polymers is 'Nylon', DuPont's original brand name.

A summary of the properties and application for BOPA is shown in Figure 2.9.

BOPE (Biaxially-oriented polyester) films are highly appreciated for their features like moisture proof and durability. These are manufactured using high grade

FLEXIBLE PACKAGING MATERIALS – BIAXIALLY-ORIENTED POLYAMIDE FILMS		
Material	**Properties**	**Applications**
BOPA film (Also known as Nylon or BOPN)	• Mechanical strength • High heat distortion temperature • High flexibility and toughness • Good barrier to oxygen, chemicals and aroma substances • High transparency	• Packaging of perishable foods • Packaging of fatty and oily foods • Packaging of foods sensitive to oxygen, such as meat, processed meat, smoked fish, cheese and dairy products, semi-finished (ready) meals • Agricultural products packaging • Packaging of medical products

Figure 2.9 Properties and applications for BOPA flexible packaging films

materials and advanced technology.

PVC films provide a range of twist wrapper, shrink labels, cosmetic shrink labels, flexible packaging materials and multi-purpose pvc bags. PVC twist wrappers come in a bright and quality construction finish and can be provided in widths from 10 to 1500mm and thickness of 20 to 100mic. Further, they can be provided in transparent or color finishes.

BOPS (Biaxially-oriented polystyrene) is hard, stiff, brittle, and crystal clear material which is readily expanded with gases to make expanded PS. It provides poor solvent resistance (can be solvent bonded) and poor overall barrier properties and is most commonly found in the form of expanded PS label stock or used for envelope windows.

LAMINATES, BARRIER MATERIALS AND SEALABILITY

There is no perfect, universal flexible packaging

material, which means that many materials used for flexible packaging involve the use of multi-layers which are created by bonding together two or more materials, whether plastic, paper or foil. An ideal laminate for a particular application assembles materials with individually desirable properties to create an optimum performance combination.

There are generally three components in a flexible packaging bonded or laminated structure – the exterior, the barrier, and the sealant. The exterior layer is the print surface. The second component is the barrier layer (or layers) which provides protection based on the product being packaged, the desired shelf life and the storage and distribution conditions required.

The sealant layer is a material that will adhere to itself, or to another film when heat and pressure are applied to produce hermetic seals that prevent gases from penetrating through the seals into the package. It is typically applied to the inside layer of a multi-layer structure on the side that comes in contact with the product. There are a number of heat-sealing materials that the converter can use. These are shown in Figure 2.10.

In flexible packaging the term barrier is most commonly used to describe the ability of a material to stop or retard the passage of atmospheric gases, water vapor, and volatile flavor and aroma ingredients. A barrier material, as earlier described, is one that has been designed to prevent, to a specified degree, the penetrations of water, oils, water vapor, or certain gases, as desired. Barrier materials may serve to exclude or retain such elements without or within a package.

There are many variations in the properties of substrates from different manufacturers. Substrates with an improved barrier are generally more expensive and more difficult to obtain. So, it is important not to over-specify the barrier required, just to play safe.

However, unless there is good seal integrity there is no point in spending money on a barrier film. Whether the sealant is a coextruded heat seal layer, a coating, cold seal, CPP or PE, it must hermetically seal the inside of the pack from the outside.

As narrower web printers move into flexible packaging they must understand barrier and sealant layers and make sure they understand how the film will be sealed, and which side seals to which side, so

FLEXIBLE PACKAGING – COMMON HEAT-SEAL MATERIALS		
Material	Properties	Performance
IONOMER (e.g. Surlyn)	Seals through contaminants	HIGH PERFORMANCE ↑
METALLOCENE POLYETHYLENE	Fast low-temperature seal	
LLDPE	Good hot tack, tough, wide seal temperature range	
PE/EVA	Soft film, low temperature seal	Cost and bond quality
MEDIUM DENSITY PE	Stiffer, better barrier	
CAST PP	Stands higher temperatures	↓
LDPE	Lowest cost	LOW PERFORMANCE

Figure 2.10 Common heat seal materials used in flexible packaging

that they can purchase the appropriate films. Most flexible packaging films are heat sealed, but not all sealant layers are compatible. Generally, sealant layers will always be compatible with themselves.

But do not assume that side A will seal to side B. Ensure to understand how the film will form into the package, and where there is an A to B seal make sure the two sides will actually seal, and there is no ink or lacquer preventing this.

When combining two or more films to make a laminate (Figure 2.11), the actual barrier performance created will be a combination of the barrier of the top and bottom (or more) layers. This is a complex issue as the converter actually needs to have the equipment to measure the above properties. Ideally, the aim would be to work with one film that provides the necessary barrier, with the other being used just to protect the inks.

UPPER LAMINATE

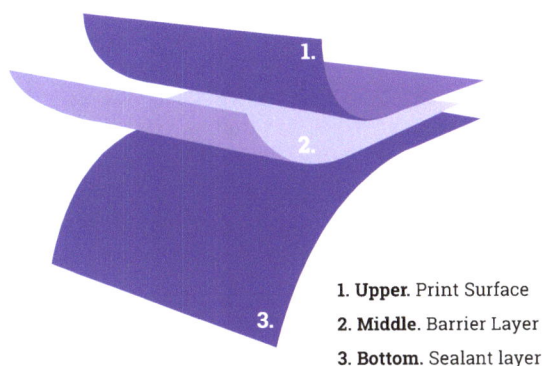

1. **Upper.** Print Surface
2. **Middle.** Barrier Layer
3. **Bottom.** Sealant layer

Figure 2.11 Basic structure of a flexible packaging laminate

With regard to laminates, it should perhaps be noted that ever greater importance is being placed on the development of sustainable packaging. Some of the key flexible packaging producers are working extensively to protect the environment through producing laminates which do not contain aluminum foil, but are made from the mono polymer family. Such laminates can be easily recycled and better fit in to today's more circular economy.

Lamination can be undertaken using either extrusion coating or adhesive laminating.

Extrusion laminating is a process in which layers of multi-layer packaging materials are laminated to each other by extruding a thin layer of molten synthetic resin, like polyethylene (PE), between the layers.

Adhesive laminating is a process in which individual layers of multi-layer packaging materials are laminated to each other with an adhesive.

As previously discussed, a substantial barrier can also be achieved through coatings or metalization. Film manufacturers have been applying these for years. However, converters do have the opportunity to apply their own barrier to film. Originally, these coatings were the same as those applied by the main suppliers of coated film. Now however, there is a newer type of coating available which allows gravure and standard flexo converters to achieve the highest

What is flexible packaging used for?

Product and commercial considerations	Technical Considerations
❏ Type/nature of the product	❏ Designation of the pack (type, style, if known)
❏ Quantity/weight/ volume to be packed	❏ Pack raw material(s): grade, quality, weight
❏ Critical attributes/ characteristics	❏ Construction of pack, if known
❏ Packing method and conditions	❏ Relevant dimensions and tolerances
❏ Product and barrier protection requirements – physical, climatic, biological, migration, security, etc.	❏ Special features/ properties/ accessories
	❏ Graphic design/ printing
❏ Quantity required	❏ Coefficient of slip
❏ Reel winding and print position	❏ Packaging line speed, sealing type and dwell
❏ Print quality, certification, barcode readability	❏ Applicable standards and test methods
❏ Pricing	

Figure 2.12 Factors to be considered in the selection of packaging materials

levels of barrier in a flexible coating that is not subject to flex cracking. These are based on mica or silica particles that provide a difficult path for the water vapor or oxygen particles.

However, a word of caution. If the coating is undertaken in-house, the converter then becomes responsible for the barrier and the protection of the food in the package, and there is a risk of very high claim levels if the film underperforms.

GETTING STARTED

One of the first challenges that the flexible packaging converter will be faced with is deciding what paper, film or films, metalized film, etc., to use or include in a flexible packaging structure with barrier performance.

Key criteria to be agreed for the selection of flexible packaging substrates

Barrier migration performance	Does the material need to protect the product from external influences; light, UV rays, gas, oxygen, moisture? What are the permeability characteristics of substrate?
Product migration performance	Does the material need to provide protection from the product inside the pack; oil, fat, moisture, chemicals, soaps, detergent? Are anti-mold or anti-fungus papers required?
Packaging performance	Does the material need to provide physical protection against puncturing, tearing, bursting, rubbing, scuffing? Is the product (solid or hard shape) likely to damage the packaging material? How will the pack be transported? What tolerances are necessary?
Sealability performance	Does the material need to be sealed? What sealing temperature will be required? What will be the dwell time on the sealing line? How will the thickness of the material affect sealability?
Print quality performance	Printability of substrate. What print quality is required? Barcode readability requirements? Rub or scuff resistance? Print position? Over-varnish or barrier coating requirements? What print and usage tests are to be undertaken?
Packaging line performance	Will the material give the required performance on the packaging line? Are there issues of static? What co-efficient of slip is required? Also refer back to sealing performance. Pack type: pouch, bag, sachet, etc.? Are there specific construction requirements?
Warehouse, transportation and storage	Does the substrate meet all the handling, storage and transportation requirements? How long will the product and pack be stored – days, weeks, months? What temperatures are involved: cool, cold, freezer? How are goods transported?
Shelf life performance and requirements	Does the packaged product meet shelf-life and consumer requirements. How will it be displayed? Stand-up, lay flat, stacked? Barcode readability? Is the printed message properly visible on the shelf?

Figure 2.13 Key criteria to be agreed for the selection of flexible packaging substrates

Plenty of polymers, barrier materials, coatings and sealants are available and in common use in the industry. A basic understanding of the material choices available (set out in the preceding pages) and the requirements of different product, commercial and technical considerations can undoubtedly simplify the selection process. These can perhaps best be summarized in the table in Figure 2.12.

If it helps to simplify decision making then Figure 2.12 can be used in the form of a tick box to ensure that all the essential elements in materials selection have been identified for a particular job, then specified and quantified. But do not over-specify. That will quickly lead to increased costs and being uncompetitive.

On the other hand it is essential to be sure to include all the elements that are required for successful substrate, pack, print, packing line, quality and end-use performance. A failure to include important items can lead to failure at one or more stages in the whole integrated flow line process. The aim should be to ask the right questions and agree the key specification requirements (Figure 2.13) with the customer – not only what matters, but also what is known to work. To a large degree this comes down to experience.

The aim should be to agree all the key materials specification requirements with the customer on what has to be delivered at an appropriate and acceptable cost. It is important to understand that the value of a substrate has to be set in the context of the end use. High-value goods with possible high failure costs can perhaps justify a more costly, higher-performance packaging material so as to guarantee a secure and fully functional delivery.

Lower-value commodity type goods may enable the selection of a lower pack performance and, where consumer or industrial end customer switching costs are small to non-existent, a critical need for cost efficiency may drive the selection to a lower package performance and overall cost-in-use.

The correct substrate specifications and selection will undoubtedly ensure a happy customer. Poor or wrong substrate selection can adversely affect the supply relationship, and may result in high wastage, rejected goods, higher costs and even non-payment.

No flexible packaging supplier can guarantee that its package is suitable for every type of product under all possible usage conditions. However, if the product manufacturer and the packaging materials supplier, ink and coating manufacturer and the converter all effectively collaborate, they usually can discover a successful solution. It is therefore crucial for all parties to maintain open communication lines and have a clear understanding of accountability for different aspects of product performance, quality, safety, standards and the environment.

The impact of packaging waste on the environment can be minimized by prudently selecting materials, following EPA guidelines.

Chapter 3

———

Types of flexible packaging and special constructions

———

In Chapter 1 it was stated that flexible packaging can perhaps best be defined as packaging that is designed to hold products, goods, solids, liquids, creams, pastes or powders and which is flexible in format, has no defined fixed shape and can be readily or easily changed. Having said that, there are many factors that need to be taken into account when designing the different types of flexible packaging – including physical specifications, barrier properties, print quality and performance, filling and sealing requirements, brand visibility, handling, storage and display.

———

How are products the pack will contain to be inserted? What protection is required? Will the packs have to lay flat? Will they need to stand up on a supermarket shelf? How will they be opened or emptied? Do they require a re-closable feature? How are the packs to be sealed or closed? Do they require an easy carry handle or feature? What about hanging holes? How will they be transported? The list of requirements or specifications today is almost unlimited, particularly when looking at the multitude of added-value possibilities that are now feasible.

There is little doubt that providing added-value features can significantly enhance the value to a customer's flexible packaging portfolio. Such features can include flip-top caps, pumps, gussets, squeezability, easy-open features, notch options, dispensing possibilities, pre-made straw holes, self-heating pouches, aseptic pouches, child-resistance, tamper-evidence, pourable spouts, zippers, and tear-off pull tabs. The list can go on.

With such a wide range of products and goods

being packaged, the range of flexible packaging types and solutions is of necessity extremely diverse and will include:

- Roll-fed film
- Wrappers
- Pillow pouches and bags
- Quattro seal bags
- Sacks
- Diaper bags
- Packets
- Sachets
- Gusseted bags
- Stand-up pouches
- Lidding
- Multi-layered structures
- Zippered, pourable, spouted and re-closable packs

In many cases the types of flexible packaging will overlap: pillow sachets, pillow pouches and pillow bags are all essentially the same, the difference comes in the

Pillow bag

Gusseted bag

Stand-up pouch

Tetrahedral bag

Quattro seal bag

Quattro seal bag with back seal

Sachet

Diaper bag

Ponytail bag

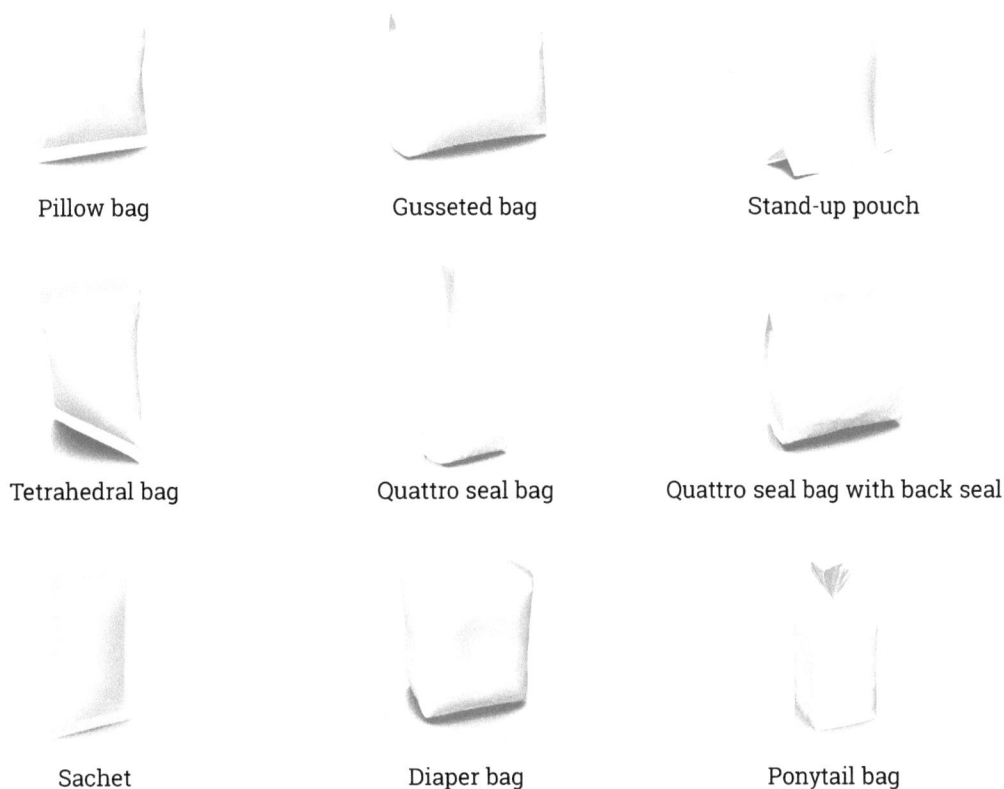

Figure 3.1 Image examples of common flexible packaging types. Source: Esko

size of the pack. Pillow sachets are generally quite small and can be printed and produced on narrow- and mid-web press, while pillow bags (depending on size) are more likely to be produced on mid- and wider web presses. Sacks and other larger sizes are not the kind of products capable of being produced on narrow- or even mid-web presses.

Examples of some of the more common of the flexible packaging products and types can be seen in Figure 3.1.

More recent developments in flexible packaging have included the evolution of spouted packages. Spouted pouches are re-closable, which makes them a good match for grab-and-go products. The pouches can even be made with die-cut handles for better portability. In addition to being customizable,

re-closable, and portable, spouted pouches offer great flexibility in the types of products they can hold, including food, beverage, condiments, dry mixes, pet food, granulates, and powders.

Look worldwide, and there are many hundreds of examples of innovation in flexible packaging to be found. Each one starts from an idea; meat should stay fresher longer, shipping costs should be lower, there should be a guarantee that no noxious substances will seep into, or out, of the product, and medicines should be safer for the consumer. Creativity and innovation have long been at the forefront of the industry. Something that label converters have also long been good at.

In an educational book of this kind, it is not going to be possible to describe every conceivable format, construction and type of flexible packaging. A more achievable aim is to cover common and popular formats perhaps most suitable for label converters looking to expand their product portfolio, bearing in mind that the aim of the pack is to:

- Contain the product
- Protect the product
- Protect the user
- Inform the consumer
- Provide for ease of movement, handling and transit
- Offer a convenient format for handling, storage and usage by the consumer.

To aid the reader's understanding, the more common and popular flexible packaging types and formats – particularly those that might be of interest to the narrow- and mid-web converter – are each set out in the remainder of this chapter.

WRAPPERS AND WRAPPING

Of all the flexible packaging solutions it is likely to be Twist wrapping that is one of the most easily recognizable – certainly by children worldwide. Its major advantage is that the ends can be easily un-wrapped or un-twisted to enable the consumption of small pre-formed confectionery items such as toffees, chocolate balls, and other types of candy products.

Most commonly made from cellophane, twist wrapping cellophane has all the required properties – good printability, strength, good twist performance. Newer polymer films are less costly but are prone to static build-up and consequent production problems. Twist wrap papers require a high twist packaging performance, but have excellent stretch characteristics, good strength and printability, a variety of opacity levels and are anti-static.

Other materials used for twist wrapping confectionery products include PVC, foil, and waxed twist paper. Caramels are usually wrapped in cellophane using a twist application due to the release properties of cellophane film. Caramels will stick to the wrapper if not wrapped in a coated cellophane film.

Twist wrapping machinery can wrap and twist shapes that are square, oblong, cubed and elongated in shape, as can be seen in Figure 3.2.

Pre-printed wrappers and decorative wrapping bands are an effective way to provide

Figure 3.2 Twist wrapping of confectionery and candies. Source: Esko

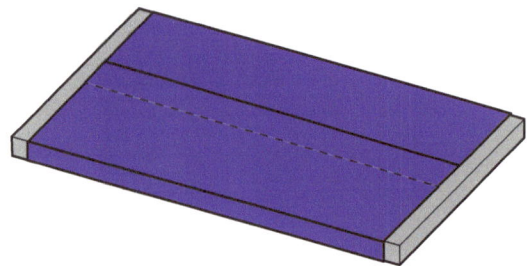

Figure 3.3 Flow wrapping of a chocolate bar

pack visibility and identity: to get the brand message across and end engage the consumer. In many cases products are completely over wrapped in paper, foil or film in a process called **Flow wrapping** – a horizontal packaging technique used for wrapping both single and multi-pack applications with a continuous flow of wrapping material – so as to form an airtight seal (see Figure 3.3).

Flow wrapping is used extensively throughout a number of industries, especially confectionery. The very nature of the flow wrapper means that the whole operation is quite fast. It lends itself to products such as chocolate (see Figure 3.4) and cookie bars, where the item to be packaged is uniform in size and shape and is a discrete piece.

Flow wrapping can be applied to various sized products, but it is most commonly used for smaller items that are either sold individually, or as part of a collection of items in a larger package. The wrapping may be clear or printed to meet different brand and display needs.

The use of flow wrapping, sleeve wrapping, food

Figure 3.4 Flow wrapped confectionery. Source: Labels & Labeling

wraps and confectionery wraps have undoubtedly become one of the key optimum solutions for the growing market for all kinds of snacks, cereal bars, biscuits, cookies, energy bars, ice cream bars, popsicles, muffins, and on-the-go products.

Films used for confectionery products wrapping cover the full range of base materials available and are very much dependent on the brand owners' desired retail requirements. These range from highly decorative basic polyethylene bags, to laminated polypropylenes and metalized materials that lend themselves to outer bags, inner wraps and provide shelf-life for ingredients such as nuts.

Non-food flow wrapping is an extremely versatile packing method that is adaptable for use with many products, including flow wrap for the encapsulation of irregular sized or solid products in clear or printed film e.g. dry, solid products such as sponges, cutlery, cloths, razor blades, soap bars, and even medical devices, hardware and industrial components.

Flow wrapped items can be presented in either a clear or printed film. Clear flow wrapping is an effective way to package small items inside larger packages, for example free gifts, whilst printed, full-color flow wrapping is a great way to promote promotional items to their best advantage. If required, holes or slot cut outs can be created on line to enable flow wrapped packs to be hung on point of sale hangers

Flow wrapping has also become increasingly popular for the packaging of promotional items, such as individual portion samples of food or beauty

products for door drops, or free gifts for packaging inside cereal boxes.

Sleeve wrapping is when single or multiples of the same products are wrapped in paper or film, with both ends open. Common uses for sleeve wrapping are for soft drinks bottles or cans (see Figure 3.5), as well as where multiple units are wrapped together.

Soap Wrapping. As soaps and detergents are chemically prepared they need to be wrapped in good quality soap wrapping anti-fungus and anti-mold treated paper for proper handling during transportation and also to protect the consumer from coming into direct contact with strong substances that are in the soaps. The ingredients from which the soaps are manufactured are mentioned on the wrapper; in addition, packaging plays a vital role in the branding and promotion of a particular product.

POUCHES

There are numerous sizes, formats, styles and shapes of flexible packaging pouches that are now available in the market place and which can be produced by many label converters. Essentially, basic pouches can be quite simply described as small bags that are most usually constructed by sealing one or two flat sheets along the required edges. There is generally no clear distinction between a pouch and a sachet other than a common understanding that a sachet is much smaller.

The most widely used of all the flexible packaging pouch products are pillow pouches and stand-up pouches, with the former expected to achieve the highest gains due to their increasing acceptance and much wider usage in the food, beverage and dairy industries. Low cost, high sealing ability and cost efficient transportation are some of the key properties of pouches that are positively influencing increasing product penetration.

However, there are many other types of pouches in wide usage, from two, three and four-sided pouches, re-closable pouches, zipper pouches, pourable pouches, spouted pouches, retort pouches and gusseted pouches. The most commonly used of the pouch types are described in more detail below.

Pillow pouches. A pillow pouch is essentially a bag or pouch in the form of a tube that is sealed at both ends (see Figure 3.6). They are most commonly produced on vertical-form-fill-seal (VFFS) machines (which will be described in Chapter 6) and are

Figure 3.5 Sleeve wrapped cans. Source: OPM Group

Figure 3.7 An illustration of a gusseted pouch. Source: Esko

Figure 3.6 A simplified drawing of a pillow pouch.
Source: Esko

characterized by seals across the top and bottom of the pack and a longitudinal seal going down the center of one of the faces. This can quite clearly be seen in Figure 3.6.

Pillow pouches are used for the packaging of a wide range of food liquid products, pasteurized or steralized, including milk, fruit juices, cream, sauces, soups, water, etc., in sizes containing from 30ml to five liters, and running at filling and sealing speeds up to 200 pouches per minute.

Non-food applications for pillow pouches are as diverse as liquid detergents, distilled water, oil, washer fluids, liquid soaps, shampoos, pet foods and vegetables.

Some of the other types of pillow pouches include the **three-sided-seal pouch**. This is a pouch that is formed by folding the web material into a U-shape and then sealing the three open sides. The pouch may be made with a gusseted bottom. Three-side-seal pouches are typically made on horizontal form-fill-seal machines.

Four-side-seal pouches are pouches produced with seals along all four edges. Four-side-seal pouches can be made from a single stock or the front and back can be different stocks, as long as they have heatseal compatability. These pouches are most commonly made on multi-lane pouch-forming machines where 16 or more pouches can be placed across the width of the web.

Gusset pouches are produced to incorporate a fold in the sides and/or bottom of the pouch, allowing it to expand and provide a 'stand-up' feature when contents are inserted (see Figure 3.7). However, incorporating side gussets will increase the overall pack size, making it too wide for printing on a narrow-web press. This can be seen in Figure 3.7. Although gusseted pouches may seem to serve a similar function to stand-up pouches, their construction is quite different. Only the front face, and the back faces either side of the seam, are decorated.

Stand-up pouches are essentially a laminated film bag, typically made of plastics or a blend of plastic film and aluminum foil, that is able to stand-up on a shelf or display area (Figure 3.8). They can be decorated front and back with high quality color printing, logos, or exciting designs, so the potential to

Figure 3.8 An illustration of a stand-up pouch. Source: Esko

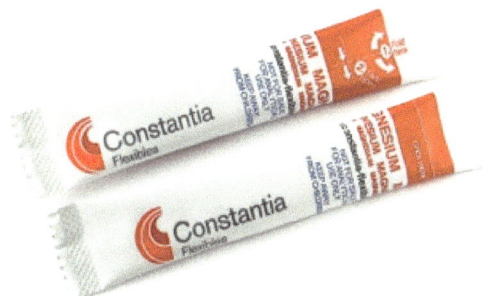

Figure 3.10 Typical stick pack construction. Source: Constantia Flexibles

Figure 3.9 Personalized digitally printed stand-up pouches

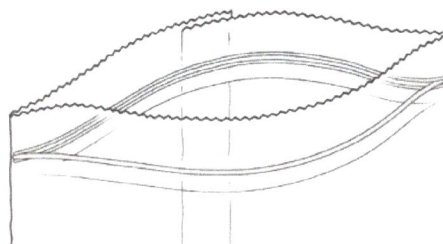

Figure 3.11 Zipper closure. Source: Esko

really make an impact on retail shelves is very high. Stand-up pouches are excellent options for both dry food packaging and for a range of liquid products.

Made from a continuous web of material, the first step after the film is printed and supplied comes in the stand-up pouch manufacturing process when the material is passed through a set of ploughs that fold a W-shape gusset into the bottom, so it can stand up.

Single or re-fill stand-up pouches in particular, are being used for an increasing range of products, from liquid detergents to beverages and lubricants – and this is expected to continue to grow rapidly. Examples of digitally printed stand up pouches can be seen in Figure 3.9.

Stick packs or stick pouches are narrow tube-shaped flexible packaging paper or laminate pouches with a fin seal running from top to bottom on the back of the pack and sealed horizontally across both ends (see Figure 3.10). They are commonly used to package single-serve powder beverage mixes such as fruit drinks, instant coffee and tea, sugar, sweetener and creamer products. They are easy to open and

simply require tearing at the top to enable the contents to be tipped or poured into bottle, cup or mug for use. Tear notches or laser scoring options for opening may also be incorporated.

The popular **retort pouch** is a flexible laminated food package that can withstand thermal processing. The choice of materials for the manufacture of retort pouches is very important. The material must have sound structural integrity and be able to withstand retort temperatures generally around 121 degrees Celsius, as well as normal handling conditions. A hermetic seal is achieved in retortable pouches by the fusion of two heat-sealable layers (such as polypropylene) to each other.

Zipper pouches consist of a flexible plastic pouch with a molded-in-place sealing device wherein a projecting rib or fin is inserted into a mating channel to effect a closure (Figure 3.11). This will add to the overall pack size. Zipper seal pouches can be

Figure 3.12 Pouch with center spout

Figure 3.14 Top and bottom seam with back flat seam

Figure 3.13 Pouch with top corner spout

repeatedly opened and closed as often as desired. The contents stay fresh and optimally protected. The Zipper pouch represents supreme convenience for the end-consumer, being convenient to handle, re-closable, and easy to carry thanks to lightweight packaging. They are easy opening, and preserve the flavor and texture of the contents.

Spouted pouches are a lightweight and convenient alternative to holding liquids, pastes or loose materials which are mostly packed in conventional bottles, canisters or buckets. Quite simply, a spout with screw cap is integrated in the top fold or corner of a pouch (Figures 3.12 and 3.13), offering both convenience and re-closability. Compared to rigid containers or boxes, spouted pouches are lighter in weight and more sustainable due to the reduced amount of material used. They can be easily filled and transported cost-efficiently as

the pouches are space saving and offer more packs per unit. Handles can be incorporated on the back of the pouch to provide easy handling, even when holding as much as five liters.

Spouted pouches offer a large communication area on each face, which is ideal for branding and eye-catching designs. High-quality printing in up to eight or ten colors and maybe special inks (e.g. metallic) ensure an attractive appearance for potential buyers at the point of sale.

SACHETS

As mentioned earlier, a sachet can be defined as a small pouch (Figure 3.14) – perhaps ideal for the printing of shorter runs by narrow- and mid-web label converters. A variety of materials can be used for sachet production, including paper, aluminum foil, paper backed foil and PET foil. Individual packs can be designed and printed in a range of colors to reinforce branding, as well as displaying all the required regulatory and product information.

Different size and shape variations of sachets can be filled with an ever-increasing range of products, including powders, tablets, capsules or liquid.

Sachets are produced as two basic types: a fin seal type which is a face-to-face seal on each side of the pack (Figure 3.15) and a pillow style which has a crimp seal on the top and bottom edges, together with a flat seam running down one side or on the back (Figure 3.16).

The majority of machines used for these types of operations, are of the vertical form fill seal type,

Figure 3.15 Face-to-face seal all around the pack, with tear notches

Figure 3.17 Sachet with hanging slot

Figure 3.16 Crimp sealing on the top and bottom edges of the sachet

Figure 3.17.

Correct selection of sachet laminates offers a number of advantages – appealing design and enhanced graphic presentation opportunities (predominately printed with flexographic printing, or digitally printed for short-runs, test marketing, etc.), and to provide long product shelf life, high puncture resistance, sterilizability, simple handling and easy opening and, importantly, a low weight, which means minimum wastage.

Applications for sachets are today as diverse as creams and gels, impregnated hand, face or spectacle wipes, liquids, pastes, sauces, creams, personal care and hygiene products, veterinary treatments, powders, granules, medical patches, haircare products, tissue wipes, and a variety of applicators in sachets, such as swab sticks.

LIDDING

Lidding films, made from aluminum lacquer coated and extrusion coated foils, foil/film, paper/film and paper/film/foil laminates are commonly used to seal and protect products such as yogurt, cream, soft spreads, cheese, jams, ready meals, seafoods, cosmetics and veterinary products that have been packed in tubs, jars, bottles, plastic or metallic trays. In addition, they provide an ideal marketing platform and an important decorative and branding function.

Most lidding films are designed to be peelable to allow easy access to the pack contents, although peel and reseal, permanent seal and also rigid lidding may

although horizontal form fill seal equipment is sometimes employed.

Sachets are often the first choice for packaging many food, medical, cosmetic and household products. The distinct advantage is that each individual sachet provides the consumer with a new, clean and fresh product each time. The aim is to promote and supply products in convenient, distinctive and affordable portions, so that the consumers get to know and like the brand. Sachets are also ideal for sampling.

For easy point-of-sale dispensing on hanging displays sachets may be produced with easy-tear notches (see Figure 3.15), and incorporate hanging holes or slots on the top edge, as shown in also be used for some applications, the later for ice creams.

Peelable lidding films are used to seal to a variety of trays and containers and are ideal for many applications including dairy products, dips, sauces, processed meats, salads, microwaveable meals, and more. Both high barrier (for extended shelf life) and non-barrier lidding materials are available depending on the requirements, as well as permanent seal (also known as weld and lock seal) and peel and reseal lidding films that can keep food on the go and prepared foods fresher for longer than products that do not have re-sealable lids.

Multi-use peel and reseal lidding films are designed for rigid or semi-rigid HDPE and PE thermoformed trays and cups and conveniently peel and reseal up to ten times to ensure safe storage and product freshness.

Permanent seal films work well for many applications including fresh cut fruits and vegetables, dairy products, sauces, and dips. High-clarity films are available in low-barrier, standard-barrier, and high-barrier structures.

Lidding may be supplied by the converter in daisy chain, die-cut and roll formats. **Daisy chain flexible lidding**, in which each lid is connected in a single continuous chain connected with a lid tab (Figure 3.18), is commonly used to seal small diameter containers. The lid rolls are unrolled on the packaging machine, sealed, and individually cut with a single cut in the tab area. A wide range of material structures are used, including aluminum, polyester and paper polyesters.

Daisy chain lidding – printed or unprinted – is widely used for portion control and single-serve packaging, from dipping sources to creamers and dressings. The processing stages of this type of lidding will be familiar to label converters, involving printing, inspection, die-cutting, waste removal, slitting and rewinding.

Die-cut flexible lidding is pre-cut (including a pull tab) to virtually any shape or size of container by the converter. The lids, which may be aluminum, aluminum laminate, co-extrusion, heatseal lacquer or plastic materials in various thicknesses and sealing layers, are subsequently loaded in stacks on fill/seal packaging machines and dispensed as the machine runs. No cutting is required on the fill/seal machine.

Die-cut flexible lidding (Figure 3.19) is used on products within the convenience food market, such as

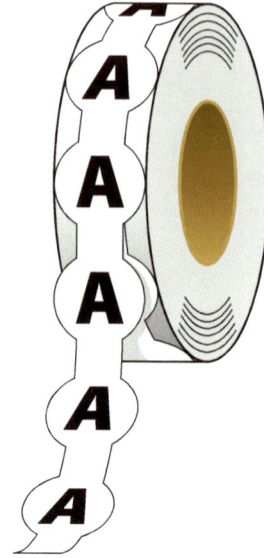

Figure 3.18 Daisy chain lidding

Figure 3.19 Die-cut lidding with pull tab

pasta, noodles, breakfast cereals, yoghurt and ready meals, as well as commonly used to seal larger diameter containers running on high-speed fill/seal packaging machines. They offer high puncture and tear resistance.

Rollstock flexible lidding (Figure 3.20) is provided in rolls (web) and then cut on the packaging machine. Materials are available for all application requirements from basic structures to sophisticated high-barrier constructions, including laminate, co-extrusion and heat-seal lacquer technologies. Rollstock lidding is most commonly used to seal containers running on fill/seal, and form/fill/seal packaging machines.

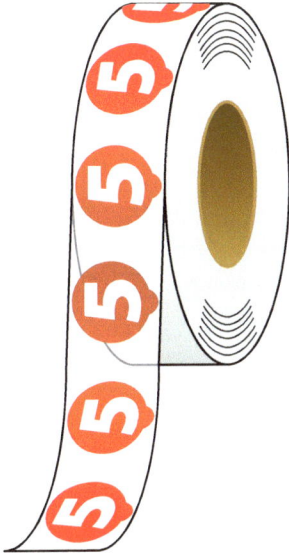

Figure 3.20 Rollstock lidding

DESIGN AND CONSTRUCTION

As can be seen from the various flexible packaging images, there are many different ways of constructing bags, pouches, sachets and lidding, many of which are suitable for narrow- and mid-web converting. A key constraining requirement for the label converter is the press web width in relation to the lay-flat total open pack size. For narrow-web presses this may only be one pack width across the web (typically around 310mm - 330mm). Mid-web presses may achieve two packs across the web with smaller packs. It should be noted that the incorporation of zipper and reseal features will have an impact on the pack design and reduce the available print/promotional area.

To better understand the structural, design and press/web width requirements of the various flexible packaging constructions the next chapter will incorporate a series of structural design templates provided by Esko, and examining some of the key design, pre-press, printing, converting and finishing requirements that the narrow- and mid-web label converter needs to understand.

Chapter 4

———

Pre-press, printing and converting

———

Flexible packaging was best defined earlier in the book as packaging that is designed to hold products, goods, solids, liquids, pastes, creams or powders and which is flexible in format, has no defined fixed shape and can be readily or easily changed.

———

Having said that, flexible packaging still needs to be designed to a specific size, and origination produced for a particular type or style of pack. Is the pack a pouch, sachet, bag or wrapper? What are its physical dimensions and tolerances? Are there any special requirements or features to be introduced, such as tear notches, hanging holes or slots, gussets, re-closure, spouts or zippers? Are there any applicable standards and test methods to be adhered to? What substrates and printing process are being used? How are the packs being sealed? If heat is used will this have an impact on the inks used?

All of these factors will need to be included in the customer order requirements and job specifications that will form the order entry and workflow for the job. Depending on the printer or converter operation, this may automate pre-press stages or become the first stage of design and origination, followed by proofing, printing and converting as required.

An area designated for date, batch or barcoding may also be required, and the client may ask for 3D visuals or physical pack prototypes of the packaging designs before finalizing the design and graphics stages. It is also important when designing the printing of packaging film to have bag drawings from the machine supplier, which will indicate exactly where printing is possible and where it is not.

For a label converter moving to the design and printing of flexible packaging many stages will be very familiar. Creating the physical size and shape; on-screen design and formatting; positioning of text, barcodes and graphics; physical trim sizes; the placing of register marks; slitting; step-and-repeat.

Creation of artwork. Certainly, label converters already using Esko label software and systems should have little problem in moving to Esko flexible packaging software. 'Studio' for flexible packaging for example, has unique design tools that enable the converter to create 3D flexible packaging shapes and packaging artwork within a matter of minutes.

A pack structure or shape can be selected from the bag, pouch, pillow, sachet, or other images in the Esko Online Shapes Store (see examples of shapes in Figure 4.1). Once selected, artwork can then be applied to that structure. Studio can add filling machine specifications, add air and liquid filling requirements, or insert a geometric shape to mimic biscuit, ice cream or other product packs. Artwork instantly appears in 3D in the Studio window in Illustrator, ArtPro or PackEdge using built-in filling and

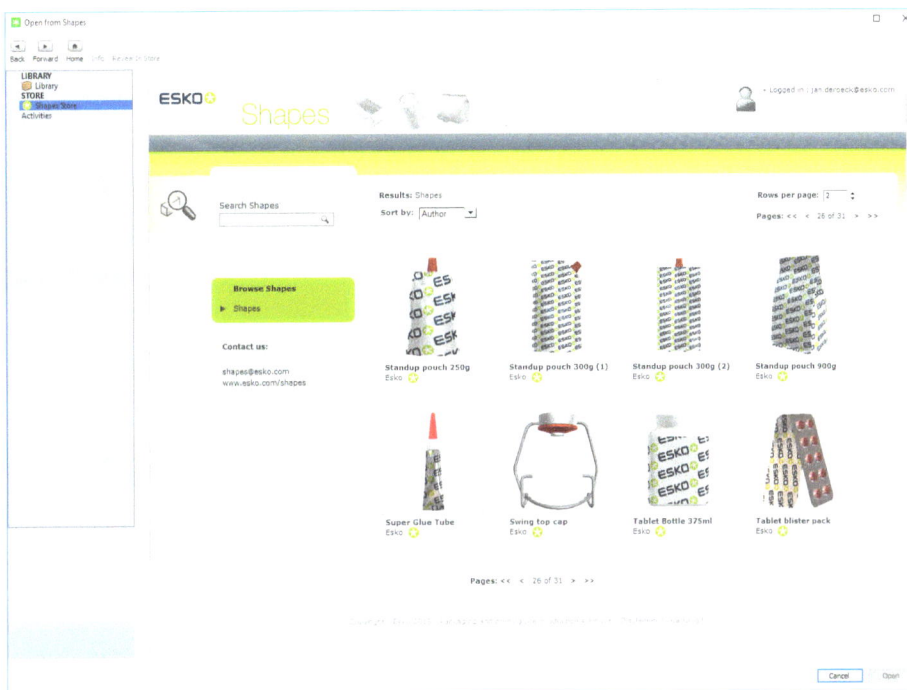

Figure 4.1 A sample page from the Esko Shapes Store showing different types of pouches

substrate knowledge.

Studio Designer enables a 3D pack preview, with all text and artwork graphics, to be spun around and looked at from any angle as if being held in the hand. Make a change to the artwork and it can immediately be seen on the 3D pack.

The operator can also see instantly whether seals or gussets will obstruct design elements. If the designs are opened in Studio Visualizer it is possible to check the opaque white backing and see any metallic inks and sealing reliefs. Visualizer can even show how the backside of the pack will appear. In addition, brand owners can see upfront how the brand will look, both in-store and when compared with the competition.

Realistic real-time images can also simulate the various printing and finishing operations, in the correct order, and on the right substrate – whether gloss, matte or coated paper, clear or white plastic films, Pantone colors and spot colors, reverse printing, and much more.

So, let's take some examples of the flexible packaging types described in the previous chapter and show how images in the Shapes Store then appear as structural images in Studio, starting with a fairly simple Pillow Pouch or Pillow Bag construction (see Figures 4.2 and 4.3).

The exact physical dimensions, bottom, top and back sealing areas, and the area available for pack graphics can all be displayed and changed as required. With this construction the front face of the pack, and the two areas either side of the center of off-center vertical back seal are available for graphic design images, brand identity, contents, ingredients, barcode, etc., as necessary to meet legislative and brand owner requirements.

Now let's compare the Pillow Pouch construction with that of a Gusseted Bag or Pouch construction, where the side gussets will normally be unprinted. This can be seen and explained in Figures 4.4 and 4.5.

This time, the pouch needs to contain side gussets, enabling the pack to expand and contain

Figure 4.2 Pillow Pouch image

Figure 4.3 Pillow Pouch structure

Figure 4.4 Gusseted Bag or Pouch

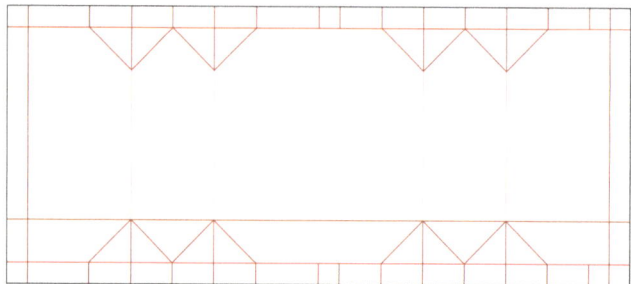

Figure 4.5 Structure of a Gusseted Bag

more content than the pillow pouch. These gusset areas on each side of the pack are clearly indicated in the structural diagram. Again, all the necessary physical dimensions, sealing areas, and the space required for pack graphics are clearly indicated. And can be changed as required.

During pouch manufacture on the form, fill and seal machine, the gussets will be pushed inwards to create the gusseted structure, the pack filled and then sealed at the top. As with the Pillow Pouch, there will be a seal area down the center or off center on the back of the pack, leaving one full face and two half faces for the graphic and text requirements. On pillow and gusseted packs the back seal can be folded (a Fin Seal) or glued (a Lap seal) in one of two ways. Left over right or right over left. These seal types will be further explained and illustrated later. Depending on how the seal is selected

the image will automatically adjust.

If we now look at the design and construction of a Stand-up Pouch or flexible bag which has top, left, right and optional bottom seals and a gusset in the bottom, the structural layout is again quite different. Different types of stand-up capability can be achieved depending on the materials used. Paper constructions and plastic construction offer different options. However, let's take just one example of a Stand-up Pouch. See Figures 4.6 and 4.7.

With this particular stand-up pouch construction, the structural layout includes a bottom double fold so that the pack can stand upright, as well as showing the side seams. There is also a larger top sealing area. The stand-up pouch has a large double front and back face area for graphics. Again all the physical dimensions and sealing areas can be created to meet

Figure 4.6 A Stand-up Pouch construction

Figure 4.7 The structure of a Stand-up Pouch

the specific design parameters and form, filling and sealing machine requirements.

A further example of a pack structure that is perhaps particularly interesting to narrow- and mid-web label converters because of its simplicity and relatively small size (multiples of which can be stepped and repeated across the width of a narrow-web press) is that of the three- or four-side seal sachet, the latter as shown in Figure 4.8 and structurally in Figure 4.9.

Both the face and the reverse of this type of sealed sachet are available for graphics. They can be sealed on all four sides, or on three sides with a fold at the bottom and seals along the left, right and top sides.

It should be noted that Studio Toolkit supports both three-seal sachets with a fold and three seals, and four seal sachets.

Converters looking to produce other types of flexible packaging can find structural templates in Esko's Toolkit for Flexibles. Dimensions and content can be easily changed and there are various tools that interact with the various shapes. Esko's solutions also allow companies to reuse text and content from existing artwork.

Esko's WebCenter includes a controlled and automated process for updating existing artwork and for creating multiple variants for a single product. All stakeholders use a common database for content, automatically pulling text statements and other regulatory content into the artwork to ensure compliance.

Artwork and color management considerations. While the flexographic process used for the majority of flexible packaging is able to produce excellent results, this is only possible if the designer takes into account the limitations of the flexo process. Designs are typically printed with a combination of the CMYK process and spot colors. Small areas of text and print edges printed in CMYK can then sometimes look slightly out of register on the printed pack due to small dot registration shifts as the web runs through the press. If large areas of the artwork are to be solid color then the use of spot colors, rather than CMYK, is likely to produce better results.

Where transparent film is to be printed it needs to be backed with a solid white color in order to make the subsequent colors stand out. Printing direct on to a transparent film, without white, will generally be harder to read, and the printed product will not look

Figure 4.8 A common style of four-side seal sachet

Figure 4.9 Structure of a four-side seal sachet

very good. If a clear window area is not necessary, then a white sealant web, such as white PE or CPP, can be used to save on ink cost.

Brand color management across different substrates can therefore be critical to successful flexible package printing. A brand color will almost certainly appear different depending on the particular substrate being used, whether paper or clear, white or metalized films. Materials will need initial fingerprinting before origination can commence. Color pallets will need to be carefully managed at the design and artwork stages and color retouching will generally be required for optimum results.

Brand color management solutions today also include intuitive software that facilitates fast, accurate press-side correction of ink formulations, with the software automating the whole process of delivering absolute consistency from press-to-press, shift-to-shift, and plant-to-plant. Accurate in-line spectral measurement will additionally enable packaging converters to achieve absolute color consistency, with automated L*a*b* measurement on film, paper, or board, ensuring that all printed products are within the customers' color specifications.

When flexible packaging artwork requires vignettes or gradient screens, it should be remembered that flexo presses cannot print a zero percent screen. Moving from a very fine screen to a zero percent screen will generally create a defined hard line edge – best avoided in the artwork design. With some substrates there may also be a tendency towards dot gain with very fine screens.

Plate gaps. An important consideration when producing plates for flexible packaging is that of the printing plate 'gap.' When flexo plates are mounted on print cylinders there is always a small gap where the top and the bottom of the plate come together. This is the plate gap, which results from mounting a flat plate on a round cylinder. This gap is typically between 1.6mm (1/8″) and 3.2mm (1/16″) wide. In most cases it can be worked in to the artwork design, enabling the gap to be barely noticeable – although still there. If the design calls for a continuous solid color with random gaps – such as a reverse-out or a negative image – then a defined plate gap is noticeable. Depending on the color being printed it may be possible to use a secondary plate cylinder to print over the gap, but this will generally only work on dark colors.

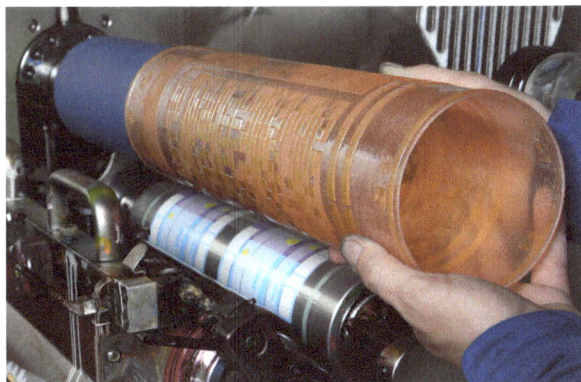

Figure 4.10 ITR sleeve mounting. Source: OPM Group

Figure 4.11 Nilpeter label and flexible packaging presses on the OPM Group shop floor

A more successful solution to the problem of plate gaps is to use in-the-round (ITR) sleeve imaging, such as DuPont Cyrel FAST ITR system. Although more expensive than conventional plate mounting, ITR systems deliver a continuous print flexo sleeve optimized for precision printing, eliminate the use of adhesive mounting tapes, offer a shorter turn-around time and improved productivity – typically running at higher speeds than flat plates, as well as working well for long runs or for short repeat runs because they eliminate the mounting process.

Imaging in position on a sleeve ensures perfect registration accuracy, eliminates butt joins, provides simplified sleeve mounting on a mandrel containing air channels (see Figure 4.10), delivers longer plate life on the press, and opens opportunities for seamless designs. The imaged sleeve simply slides over a print cylinder mandrel containing compressed air channels. When the air is withdrawn, the sleeve fits tight on the mandrel with minimal risk of plate lift.

PRINTING

Label converters looking to extend their production into flexible packaging will already be skilled in label printing technologies – whether flexo, offset, combination process, digital or even hybrid. It is not proposed to go into these printing technologies in this book. The Label Academy books on Conventional Label Printing processes and Digital Label and Package Printing amply cover these topics. The aim with this title is to look more at areas where

differences in press design, printing, inks, web handling, etc., become increasingly important when producing flexible packaging – particularly films, foils and laminates.

Certainly, flexo is the dominant printing process for both self-adhesive labels and flexible package printing process worldwide, with more than 80 percent of printers using this process. Offset is used for some flexible packaging, while digital printing, from a small base, is now growing faster than conventional press investment.

Gravure is still an important process for wide-web flexible packaging of very long runs. While some label converters may have a gravure unit on their presses, this is more likely used for coating and lamination requirements.

Flexo undoubtedly excels at printing on a range of flexible packaging materials: plastic, foil, acetate film, brown and white paper, and other materials typically used in flexible packaging. That means that to print packaging like bags, sachets, pouches, wrappers, flexible tube films and laminates, as well as self-adhesive labels and lidding, flexo is likely to be the first choice for the many applications of medium and shorter run jobs that are of interest to the label converter. The process offers fast turn-around times for plates and finished products, and can be used with a wide variety of inks, including water-based, solvent-based and UV curable.

Digital printing, or digital hybrid presses, will

increasingly come into play for ultra-short runs, multi-versions and variations, variable text or graphics and sequential coding or numbering. Certainly, emerging capabilities to print directly onto thin plastic films using digital inkjet technology are fast improving print productivity, lowering costs, and enabling not only short runs and personalization that meet changing market needs, but also offer brand owners product packaging that exudes a clean look and professional feel that has previously been difficult to replicate.

In addition, the development of HD Flexo printing now also enables converters to successfully compete with offset for quality labels and packs, and with the gravure printing process used for wide-web flexible packaging. In addition, digital plate exposure today ensures greater consistency in plate production and in flexible package printing.

As a general consideration, whatever printing process is being used, it is essential that both raw materials and finished roll handling is undertaken with care. Dirt or dust on rolls can affect print and rewind results and create possible print defects, web splits or tears. See reference under web cleaning later in this chapter. How and where incoming and finished rolls are stored can also be important. That includes temperature and or humidity levels during storage and handling.

The same comments should be applied to the press. Materials running at high speed can attract dirt and dust from the atmosphere and in and around the press. It is therefore important that presses are kept spotlessly clean to minimize such problems. This includes proper cleaning and inspection of inking rollers, impression cylinders, transfer rollers and systems to ensure that particles are not transferred to plates, cylinders, sleeves or substrates during printing.

Let's therefore concentrate on those areas of flexible package printing and converting where more particular attention and investment may be required by the label converter as the product portfolio expands and changes.

Press web width. When printing flexible packaging to be used with vertical form, fill and seal machines it is important to remember that the film, paper or foil material width must be at least twice the bag or pack width, plus the width of the longitudinal seam. Conversely, this means that the bag or pouch is half the film width, minus the width of the longitudinal seam. This can be explained in Figure 4.12.

In the example shown, the face of the pack is 146mm wide; the back of the pack is in two halves, each 73mm wide, on either side of the longitudinal seam. The seam allowance in this case is 18mm in total, with 9mm on either side of the pack overall width. This gives a total pack and substrate width for this example of 310mm, plus any additional edge trim required. This enables one pack across, say, a

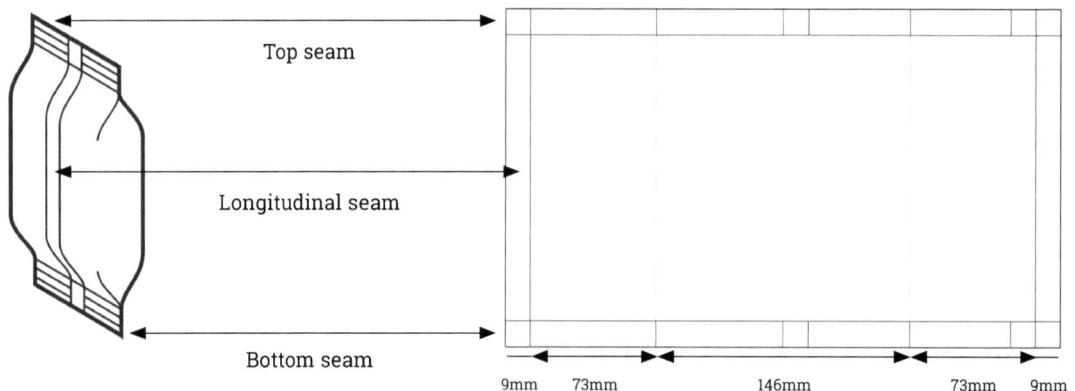

| 9mm | 73mm | | 146mm | | 73mm | 9mm |

Figure 4.12 Calculating the press web width

330mm web width press.

Even with a press width of 330mm (13") the market opportunities for a label converter to move into flexible packaging can be somewhat limited. Successful narrower web converters producing both labels and flexible packaging are more likely to be obtaining orders with flexo presses with 430mm (17") or 450mm (18"), or even wider wide web widths.

Seams. It should be noted that the longitudinal seam may be formed in one of two different ways, which affects the seam width, and press width measurements. In a straightforward overlap seam, one edge is placed over the other edge (Figure 4.13 left) and the two edges are sealed together. However, this type of seam is only possible if both the inside and the outside of the substrate are sealable, most commonly using polyethylene (PE) or polypropylene (PP), or the substrate has been coated on the outside with a seal coating. Because overlap seals require less material than fin seals, packagers are converting to lap seals in the name of sustainability, lean operations and economics.

It should be noted that where packs have an overlap seam, the exterior of the substrate at the cross seam should not be pre-treated for printing, as this make a good seal impossible.

Depending on the packaging material and the type of pack, the longitudinal seam may alternatively consist of one edge of the pack being folded over and then sealed to the inside of the other edge (Figure 4.13 right). This method of sealing is also called a fin seal. Please note that the fin seal method will require a larger pack overlap. In general the longitudinal seam

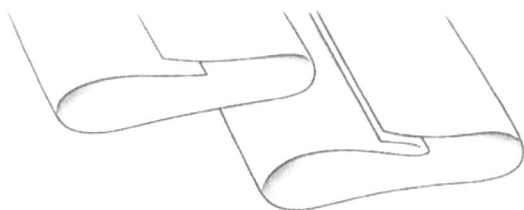

Figure 4.13 Overlap seam (left) and fold-over or fin seam (right). Illustrating the two different types of longitudinal seam overlap

overlap will be between 15mm and 25mm, depending on the seaming method.

Barcode orientation. For ease of scanning the barcode on retail packs needs to be in a clear, consistent, standard location i.e. on the rear or (towards) the base of the item. The Symbol should be printed with sufficient clear space (light margin) around it for the scanner to recognise it as a barcode and a recommended 8mm and a minimum 5mm away from any packaging seams or folds, on a flat or consistently curved surface.

On some small curved pouch surfaces, depending on the size and diameter, the barcode may be printed vertically to the curvature in order to minimize any possible problems and to achieve better results.

Registration marks. For both accurate web printing and inspection, and the subsequent form, fill and seal machine operation, a registration mark will be required to be printed on the web. A combination of the register mark on the printed web and a photo cell on the forming and filling machine will guide the web transport, with one register mark for each pack length – which in turn assures one bag length of transport per forming, filling, sealing and pack cutting machine cycle.

Tension and edge control. To ensure consistent web tension, particularly for some of the thinner gauge monofilms used in flexible packaging, it is suggested that both the press unwind and rewind are servo driven. If the tension is too high or too low it can lead to uncontrollable stretching of the film, create differences in pack length and misaligned printing, out-of-round rolls, register variations, blocking, collapsed cores, telescoped rolls, bagginess or corrugations – factors which are very important when it comes to handling, storage and downstream on the customer's form, fill and seal machinery.

Rolls that are wound too tight may also cause 'blocking', with the web layers sticking together. Such rolls also have a high residual stress or high in-wound tension, causing the filmic materials to stretch or deform when the stress is relieved as the rolls cure during storage.

Some flexible packaging films, either occurring during the coating or laminating process, or during the extrusion formation process, will have cross machine variations of thickness, which become exaggerated when the reel is wound and create high spots and ridges.

Finished roll in-wound tension is a therefore a key factor in whether a printed flexible packaging roll is of a good or poor quality and will perform well at the bag, pouch or sachet forming stage. Ideally, rolls need to be wound tighter at the core and then wound with less tightness as the roll gradually builds in diameter.

Multi-layer film laminations that often contain a barrier layer of foil utilize similar web tensions as paper-face materials and similar to the pressure-sensitive material tensions that label converters are familiar with. However, thin films that contain just a single or only two layers of film laminated together, require a press more designed specifically to convert the thin materials. Typically, this would include the following:

- An enhanced level of tension control above that required for a pressure-sensitive label press
- Lower tension servo-driven unwind and re-wind
- Utilizing a web path that has been designed for extensible materials, including idler rollers and low friction, easy turning bearings
- Optimum roll handling
- The introduction of a cold/cool UV light system
- Incorporating a chill roll or chill drum for web temperature control.

Optical web guides for optimum print edge tracking of film webs are also regarded as an important investment. This is because the film web must run through the machine as straight and simply as possible. The course of the film web is influenced by a number of factors. One factor to be considered is the printing itself. If the film material has more colors on the side that is to be on the front of the bag or pouch than it does on the side(s) next to it (the future back of the bag), the film web may tend to move sideways; the extra ink that is needed for the larger number of colors ends up making the film thicker on that side, causing uneven tension.

Rolls of film or laminate where one side of the material coming off the roll is loose or baggy while the opposite edge is tight, is said to have a baggy edge. In processing flexible packaging materials, slack areas in the web that should be flat can also be caused by bands of unequal thickness (gauge bands) in the rollstock itself.

What this means is that if the film web tends to move gradually toward one side, then the longitudinal seam will not be sealable after some time; the two sides of the film that form the tube will also not touch each other anymore. In the phase before that, the printing on both sides of the longitudinal seam will no longer match up properly, either. A dual axis servo web transport system will ensure that the press will continuously deliver the highest levels of print reproduction as well as unparalleled registration.

Automatic sideways correction using optical, ultrasonic or photoelectric web guides can be essential features for edge detection, correction and ensuring parallel rewinding. It is very important that the printed web is wound well and evenly on the core; the side of the roll must be completely flat.

Web treatment. With the wide range of film, paper, foil and laminate material used for flexible packaging the best results will be achieved by increasing the performance and surface energy of the substrate using integrated web cleaning, on-rail dual-side corona treaters, anti-static bars, and the use of chill rolls. Corona discharge is frequently used on polymer-based substrates that have low surface energy. Such substrates may have poor adhesion of inks, adhesives, lacquers and coatings.

With web treating, the surface of the film or other material to be treated is bombarded with electrons to increase the surface wettability of the substrate (generally filmic). The use of web cleaning ensures consistent corona treatment, which allows impermeable substrates to be correctly 'wetted.' This helps to ensure very good process and line print quality and excellent print registration.

Web cleaning systems are a critical piece of equipment in the modern printing, laminating and converting environment, controlling dust and contamination on the web and improving productivity. They are specifically designed to neutralize electrostatic loads and remove contaminating particles of dust attracted to reels of plastic or paper films. The presence of contaminating particles results in non-conforming products and, in particular, if the products are intended for the food, medical and pharmaceutical industries. Web cleaners guarantee to eliminate all these problems, allowing optimum print quality and impurity-free finished products.

Chill rolls. Filmic materials demand lower press, drying, and ink-cure heat levels, as excessive heat can cause them to soften or stretch. Even a difference of one to one-and-half degrees Centigrade can be critical. To minimize softening or stretching, chill rolls can be used to keep the web temperature low in addition to offering higher air flow.

The function of a chill roll or drum is to provide a uniform and consistent temperature across the chill roll and the web, using an outer and inner roll assembly. A coolant, usually water, is introduced into the inner tube, thus providing heat transfer from the outer rotating chill roll to the circulating coolant. The coolant flow rate is an important aspect of chill rolls, determining the ability of the roll to remove the required heat and temperature variation across the roll face. Sufficient flow needs to occur to ensure proper cooling.

Web inspection. Label converters with 100 percent web inspection are already well aware of the benefits such systems offer in terms of product quality and eliminating waste and errors. The printing of flexible packaging will undoubtedly benefit from the same systems and controls. Integrated with the automatic storage of inspection data and job set-up, improved back-up of job parameters/inspection data, and the availability of management inspection data, all offer the converter enhanced management information and performance, as well as contributing to enhanced profitability. It should be remembered however, that tolerances in reel splices all need to be agreed in advance with the customer

The key difference with flexible packaging when compared to label production is that with labels any waste or faults can be very easily spliced out or labels replaced. This is not possible with a flexible packaging reel. Individual pack images cannot just be replaced like a label image can.

Certainly, an optimum print inspection system will automatically detect critical defects, while eliminating false alarms, enhancing the overall workflow process. All print defects are categorized according to type with easy on-screen view, indicating the precise defect location on full repeat images.

However, an area of key importance for flexible packaging converters is the number of splices required to remove print waste in a reel. Many brand customers and filling plants will not tolerate more than a couple of spices in a roll, so if a reel ends up with multiple splices then much of the reel could be discarded as scrap – and the converter's profit margin lost. In some cases the amount of waste discarded could account for up to a quarter or one third of the printed reel. Good practise is for the converter to agree in advance with the customer what level of splices are acceptable/unacceptable.

One way to minimize such waste is to use an inspection system that calculates and provides a yield prediction for each printed reel. The system knows where the waste is located and then calculates where best to take out faults so as to provide the optimum reel yield. There may still be waste, but the best case result can be determined.

INKS, COATINGS, CURING AND LAMINATING

Flexible packaging has become popular in various industries ranging from high street consumer products right through to industrial applications. Both flexographic and gravure printing have been the traditional methods of printing used for flexible packaging printing by the majority of wide- and mid-web producers. These two methods largely share the same ink raw materials, whether solvent or water-based; hence guidelines to select raw materials for the both printing methods are common and widely understood by the larger size and larger volume producers.

Their large presses will have sophisticated servo-driven web tension and web handling controls, chill rollers or drums, web treatment, in-line coating, lacquering and laminating and hot air dryers, all designed for the handling of thin, heat-sensitive films and the required heat sealability. With many label presses, both conventional and digital, now in wider mid-web widths capable of matching some or most of these requirement, the whole market for shorter flexible packaging runs, versions, variations, personalization, sequential coding or numbering, now becomes a major opportunity.

Apart from print performance, flexible packaging films and inks may have to withstand lamination, sterilization, pasteurization and thermo-sealing, and therefore may need to be surface coated or treated. Certain aspects of flexible packaging inks therefore need to be well understood, especially where they may differ from label inks.

For example, the sealing of pouches and sachets can expose print surfaces to temperatures up to 375 degrees Fahrenheit. Inks and varnishes will therefore need to be able to withstand these kinds of temperatures. This means that where the exterior of monofilm is to be printed, then heat-resistant ink must be used.

It should also be noted that when printed flexible packaging materials are wound in rolls off the press, the print surface comes into contact with the sealant material. Consequently, inks and varnishes need to be properly cured to avoid damaging the sealability of the material. The odor of the cured inks and varnishes also needs to be considered as the odor can transfer to the sealant film and ultimately into the finished pouch.

Many pouches are widely used for food packaging. Inks and varnishes for the food market need to meet the end-use food label and packaging requirements, with low migration UV flexo inks being selected. A water-based varnish will effectively seal the inks from transferring to the seal area. Ideally, inks used for food packaging will need to be totally encapsulated. In all cases, the ink supplier should be contacted to verify they are fit for use for the specific application. The Label Academy Inks, Coatings and Varnishes book provides invaluable advice and guidance on ink technology, ink usage, proofing, legislation, tests and testing procedures and should be regarded as recommended reading.

The key challenge for the narrow-web industry as it moves into flexible packaging is the historically dominant use of the more traditional forms of UV curing, which many buyers are not prepared to countenance because of historical worries about ink component migration. Additionally, mercury lamps emit only 20 percent of energy as UV light. Some of the remaining energy is emitted as infrared light. This can cause surface areas to be cured that were not intended to be cured and over-curing of some surfaces. In addition, mercury lamps operate at quite high temperatures which can damage heat sensitive substrates, especially some thin films. This may restrict the flexible packaging materials used when operating with mercury lamps.

Using cold/cool UV curing systems reduces the UV curing temperature and can help to improve printing quality. Cold UV curing temperature can also help improve a UV lamp bulb's working lifetime. Additionally, the lower curing temperature is good for curing a UV lacquer on the thermal sensitive material. For example, even for printing base film as PET film, lower curing temperature will help improving printing accuracy. However, if the ink is not formulated for

'cool UV' systems then they may not cure as fast. A further consideration is that of the chill roller. If it is too cool then curing may again be too slow.

Laminate construction may minimize some of these issues, while UV LED based lamps have begun to replace mercury lamps for an ever widening range of UV curing applications, enabling the various challenges to be addressed.

UV LED (UltraViolet Light Emitting Diode) is fundamentally a more compact technology than traditional UV lamps – a different wavelength and intensity compared to conventional UV – enabling much more flexible form factors. The compactness of LED systems, along with the fact that shuttering and air extraction systems are no longer necessary results in a much smaller footprint for the UV curing system. Different ink chemistries are of course required. More detailed information of inks and coatings can be found in Chapter 5.

In general, inks for flexible packaging can be divided into surface printing inks and lamination inks. Inks used for surface printing must have a reasonable high gloss, excellent rub resistance, and resistance to a variety of external elements such as solvents, water, detergents, chemicals, etc., that may come into contact with the printed surface during packaging, storage, handling, display or usage. However, surface printing inks and lamination inks are slightly different, but this difference is important to note.

Surface printing inks are typically formulated with waxes that migrate to the surface of the ink film to provide rub, scuff and other resistance properties. They must also have good gloss, particularly if they are not to be overprinted. Surface printing inks will therefore typically use a protective overprint varnish to increase the surface resistance and to impart the desired C.O.F. and gloss.

Lamination inks however, are typically sealed inside a lamination structure so that the ink surface is not exposed to the outside, however the inks must still exhibit excellent adhesion to the film on which they are printed and they must not interfere with the bonding of the lamination films. Lamination is essentially the process of combining two or more substrates to form a tough packaging structure to hold and provide barrier protection for the contents, or where the surface print needs to be buried.

Lamination inks do not have the gloss

requirements of surface inks but they do require excellent bonding ability to ensure that the lamination structure isn't compromized after manufacturing. They typically do not incorporate a lot of waxes or surface active ingredients as these can often interfere with lamination bonding.

IN-LINE LAMINATION

Laminating is the process through which two or more flexible packaging webs are joined together using a bonding agent to improve the appearance and barrier properties of the substrate. The choice of the most suitable web laminating process is mainly dictated by the end-use of the product. A number of different technologies are available that cover the wide variety of applications in the food and non-food packaging industries.

Laminating machinery can be classified according to the type of bonding agent used to produce the laminates. These types are:

- **Solventless lamination**: The adhesives used do not contain solvents. They dry by chemical reaction and therefore do not requiring a drying system. This method is used widely in flexible packaging since the chemistry is relatively simple and the applications broad.
- **Wax lamination**: The adhesive is a wax or hot melt which is applied in a liquid state to one of the substrates prior to the substrates being brought together. This process allows the production of paperpaper or paper-aluminum foil laminates that are widely used for the packaging of biscuits and bakery products.
- **Dry lamination**: Is the process where the bonding agent, dissolved into a liquid (water or a solvent), is applied to one of the webs, before being evaporated in the drying oven. The adhesive coated web is laminated to the other under strong pressure and using heated rollers, which improves the bond strength of the laminate.
- **Wet lamination**: With wet lamination the adhesive is in a liquid state when the laminate and base substrates are brought together. It is commonly used to produce a paper-aluminum foil laminate that is widely used in flexible packaging and in the self-adhesive labeling industry. In the process of wet lamination the adhesive is applied to either the reverse side of

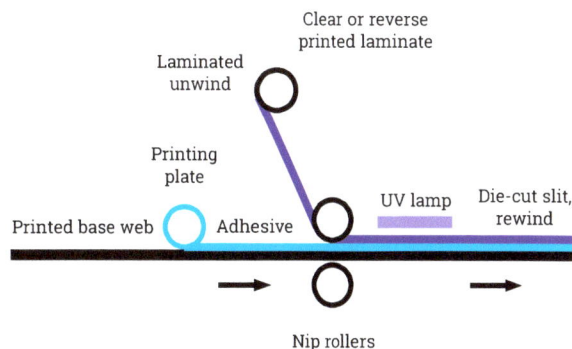

Figure 4.14 The wet lamination process

the film laminate or alternatively to the face of the printed substrate to be laminated.

The liquid adhesive is applied by a roller coating system. The two substrates, one of which has been coated with the adhesive, are then nipped to form the bond (see Figure 4.14).

In all of the laminating processes described the resulting laminated web is then rewound into a finished roll.

Label converters moving into laminating will probably already be most familiar with adhesive laminating.

It should be noted that adding an overlaminate may require the packaging equipment to run at higher temperatures and therefore a destructible bond is required between the overlaminate film and the base film structure. It is highly recommended that the overlaminate is one that has been specifically designed for flexible packaging applications.

The two main methods of applying a flexible packaging overlaminate to the web on a narrower or mid-web label press are:

- Either, a self-wound thin PET laminate can be applied on the press. This is similar to using an overlaminate on a label press. Special handling of the overlaminate on the press may be required due to aggressive adhesive and thin film.
- Or, a wet adhesive laminate using UV lamps for the curing. This will require UV lamps to be in premier state, with a curing time for the structure between 24 and 72 hours.

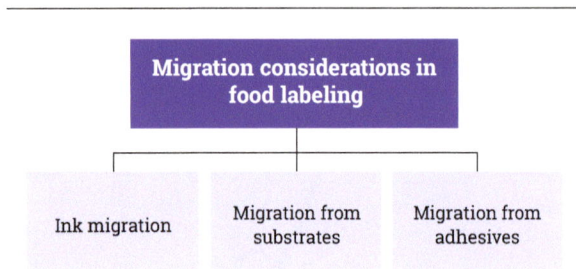

Figure 4.15 Key areas of migration to be considered in food packaging and labeling

BARRIER COATINGS

A barrier coating or sealing coat can be applied to flexible packaging webs to prevent migration of ink, adhesive, or other substances through the face material. Barrier and other functional coatings encompass materials that are coated onto substrates to provide a barrier to protect selected packaged goods.

Barrier coatings, providing barriers for food packaging requirements, may include protection against oxygen and aromas, liquid water and water vapor, oils, and grease. An effective barrier can prevent both losses from the packaged product, and penetration into the package, both of which can affect quality, and shorten product shelf life. Packaged food products are being maintained fresh longer as a result of new materials, and food processing developments. For example, O^2 scavengers are now being used that work within a sealed package to limit O^2 reaction with a food product. Combined with effective O^2 barrier packaging, food packagers have the ability to improve shelf life, preserve product appearance, and flavor, while minimizing preservative use.

Additionally, antimicrobials, while under siege, have been proven effective as additives to coatings, and packaging films, in combating food sourced illnesses. Nanotechnology is being applied to improve the gas barrier properties of coatings. In doing so, nanoclay is dispersed in barrier coatings, resulting in a platelet orientation that creates a 'torturous path' for gas molecules to traverse, yielding a very thin film, effective gas barrier.

Knowledge around migration and/or barrier protection is crucial for servicing the flexible packaging food markets. Suppliers must be able to provide the right barrier or seal for a given application and be able to support clients with 'fitness for use' testing.

PACKAGE SEALING CONSIDERATIONS

Cold Seal Packaging. Many temperature sensitive confectionery items (e.g. chocolate) are packaged and sealed using cold seal adhesives. These surface printed constructions consist of a surface printed ink and a cold seal release lacquer (CSRL), which prevents the printed ink from offsetting against the cold seal glue when in the printed roll.

Inks for cold seal packaging should not be formulated with any kind of fatty amide (erucimide) or PFE waxes as these are said to 'poison' the cold seal adhesive if left in contact with the adhesive for any length of time in the printed roll.

Heat Seal Packaging. Both surface printed inks and lamination inks can also be heat sealed. In this instance a heat sealable film is used or a heat sealable coating applied to the packaging material is used to combine two films by applying heat to achieve the seal. Sealing temperatures and pressures can vary so when formulating these types of inks, it is important to know and test the inks under the specific sealing conditions. A typical sealing specification may be 350 degrees Fahrenheit for half a second at 40p.s.i.

COEFFICIENT OF FRICTION/SLIP

Before leaving this chapter it was considered important for label converters moving in to flexible packaging to have an understanding of the role that coefficient of friction/slip can have in successful printing and subsequent packaging line performance.

Coefficient of friction (COF) is a term that is significant in the world of packaging and package printing and is critical to the optimum processing and handling of packaging from filling operations through to the consumer. By definition, COF is a measure of slip, or how one surface moves across another. The importance of this is in how a substrate moves through and across print stations, conveyor belts, on the form, fill and seal machine, or released from a mandrel or loading after processing.

It should be noted that there are two kinds of coefficient of friction: static, which is the force needed to begin movement, and kinetic, the force required to maintain movement. Generally, static COF is of greater concern for stacked or palletized items, while kinetic

COF is important when using roll films. It provides a relative indication of frictional characteristics and is routinely specified in substrates such as plastic films used by flexible packaging converters.

Films with COF values greater than 0.5 are considered non slip. COF is usually specified for a given process and adjusted by the printer with inks or varnishes as needed. Depending on the packaging application, a high or low COF may be desired. But why does it matter?

In flexible packaging applications where a film web is pulled over a forming collar, such as with vertical form, fill and seal equipment (explained in the next chapter), a low COF is considered to be favorable. If too high, the flexible film may bind when sliding over the filling collars. However, if the static COF is too low, problems in maintaining stability in stacks could result, as well as difficulty pulling materials through automatic processing machinery. The importance of COF requirements in different filling and packaging applications can be summarized as follows:
- In VFFS (vertical form fill and seal) systems, too much friction of the sealant side of the film can cause poor film feeding over metal forming collars and inconsistent package sizes
- In HFFS (horizontal form fill and seal) systems, too much friction of the sealant side of the film can lead to film dragging or jamming as it passes over metal plates
- In either system, too much friction can result in lateral slipping that leads to poor seals (leakers)
- Too little friction on the outside can cause packs to slip or fall-off-of inclined conveyor belts
- Too much friction on the outside can slow packages' progress down delivery chutes
- Too little friction on the outside can result in packages sliding off of stacks or pallets.

Controlling COF gives the form, fill and seal processors the ability to optimize performance and avoid problems in forming, filling, transporting, and storing of the finished packages.

Getting the COF right is mostly a process of qualifying a packaging product and then supplying this product at a consistent COF specification based on a given instrument and testing procedure. It is important to remember that many converting variables cause COF to vary. COF can be affected by a number of factors, including antiblock additives, corona treatment, inks, varnishes, adhesives, application viscosity, coating weight, drying conditions, and environmental humidity, while in energy-cure systems, COF is affected by the degree of cure. Running to proven standards will ensure low waste and fewer field problems and returns.

In packaging films, the traditional technology to lower COF is to incorporate fatty amide, waxes, and silicones, and add silica particles to increase COF. Newer technology uses more stable proprietary non-migratory slip packages.

Customized COF's are achieved by adding a 'slip agent' to a film resin during production. The traditional approach to reducing COF involves adding a compound that is incompatible with the film resin, and will migrate to the surface of the film over time. Non-migratory slip agents offer benefits in the area of thermal stability and consistency, but can affect film clarity.

Label converters moving into flexible packaging production should consider investing in COF testing equipment, of which there are two common basic types — sliding plane and incline.

CONFIGURING A LABEL AND PACKAGE PRINTING PRESS

What can be seen from the information in the preceding pages is that a narrow/mid-web press designed for label and package printing will need to meet certain web handling, control, print and finishing requirements. In particular, the press will need to be able to print and cure/dry all the necessary types of inks, varnishes and adhesives (see more information in the next chapter), whether on supported or unsupported films, and in web widths increasingly offered by key press manufacturers of 420mm, 430mm, 450mm, 480mm, 500mm, 520mm, 620mm and even going up to 750mm wide.

Stages in the flexible packaging printing and converting process are:
- Printing
- Coating (varnishes, primers)
- Laminating
- Inspection/slitting/rewinding
- Forming, filling and seaming

Factors to be considered include register control, web tension, rewind tension, methods of cure/drying,

MULTILAYER CONVERTED WEBS FOR FLEXIBLE PACKAGING

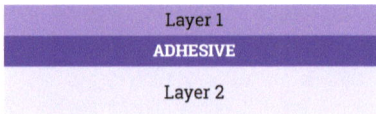

Layer 1
ADHESIVE
Layer 2

Two layers laminated products:

DUPLEX

Layer 1
ADHESIVE
Layer 2
ADHESIVE
Layer 3

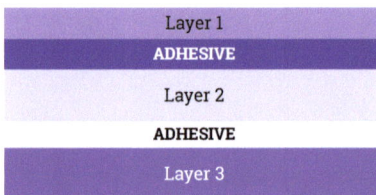

Three layers laminated products:

TRIPLEX

Figure 4.16 Construction of multilayer converted webs for flexible packaging. Source: Bobst.

heat control, smell, and special applications such as heat-seal and cold-seal. In terms of register and web tension control the converter will likely need to consider a short web path, a free running counter pressure roller, web transport by chill drum, low tension and paper tension rewind.

In terms of curing, the flexible packaging printing and converting press has to be able to cure/dry using all types of flexible packaging inks, varnishes and adhesives:

- UV
  - Mercury
  - LED (hybrid or not)
  - Nitrogen
- Hot air
  - Water-based
  - Solvent-based (explosion proof)
- Infra-red
- EB

New generations of label/flexible packaging press may be hybrid machines, have application modules on rails and incorporate intelligent web transport systems.

Where lamination of flexible packaging webs is to

EXAMPLES OF DUPLEX DRY LAMINATION OF TWO FLEXIBLE PACKAGING SUBSTRATES		
Packaging	Compound	Adhesive
Snacks, Pasta	Film/Film, Film/Foil	Solventless
Spices, Granulate for soups	Film/Foil	Solventless
Frozen food	Paper/Film, Paper/Foil	Solvent-based
Biscuits	Film/Foil	Solvent-based
Meat, Sausages, Cheese, Washing powder, Fresh fruit, Fish	Film/Film	Solvent-based / Solventless
Pharma products	Film/Film, Film/Foil	Solvent-based / Solventless
Blister for Pharma products	Film/Film, Film/Foil	Water-based / Solvent-based
Fruit juice	Film/Paper	Solvent-based
Cosmetic products, Blister for precooked food, Cremes	Film/Foil	Water-based

Figure 4.17 Examples of Duplex dry lamination of two flexible packaging substrates. Source: Bobst.

be carries out this may be undertaken all in one pass (print/UV laminate/slit/rewind). Alternatively, the web may be printed and laminated in one pass, stored for 8 to 24 hours and then slit and rewound. Another option is to print, to laminate, store for 24 hours and then again slit and rewind.

Laminated flexible packaging webs may be two layer laminated products (Duplex webs) or three layer laminated products (Triplex). This can be seen in Figure 4.16.

Examples of a typical two layer Duplex dry lamination flexible packageing web showing the packaging application, the Duplex construction and the laminating adhesive used can be seen in Figure 4.17.

Chapter 5

—————

Inks, coatings, curing and laminating

—————

So far in this book it has been possible to cover a number of key areas in relation to the production of flexible packaging: materials, pre-press, and the printing processes – both conventional and digital. What has not so far been discussed in any detail is the nature and role of inks, coatings, curing, varnishing and laminating used in the narrow-to-mid-web flexible packaging sector.

—————

In particular, the use of UV-curable inks and the overall requirements of UV-curable food contact material (FCM) inks in terms of:

- Regulations
- Performance
- Economics

These areas will be examined during this particular chapter.

A number of previous chapters have so far referred to why UV-curable inks are now being used for flexible packaging: shorter runs, reduced delivery times, personalization, differentiation, etc. All factors which point to why narrow-to-mid-web flexible packaging technology presses are often the best solution for such applications. This can be amplified in the following diagram (Figure. 5.1)

This diagram provides a simplified illustration of what is happening nowadays in the market? Wide webs are primarily best for jobs of, say, 100,000 square metres, run at very high speeds. However, with the trend to shorter runs, personalization, differentiation, etc., a better suited technology for the short-run of less than 5,000 square meters can be

WHY PRINT PACKAGING IN NARROW WEB?

What is the typical run length / job?

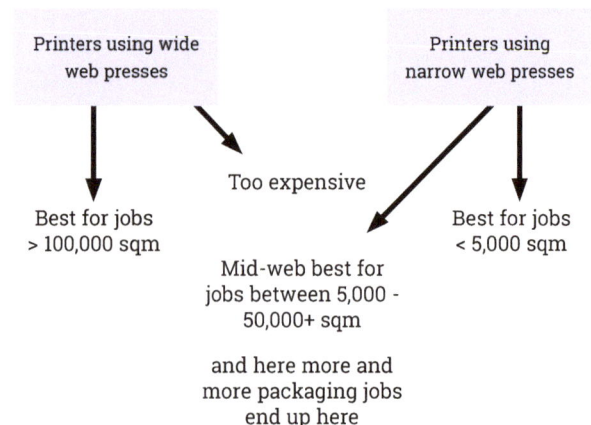

Printers using wide web presses

Printers using narrow web presses

Too expensive

Best for jobs > 100,000 sqm

Mid-web best for jobs between 5,000 - 50,000+ sqm

and here more and more packaging jobs end up here

Best for jobs < 5,000 sqm

Figure 5.1 Why print flexible packaging on narrow- or mid-web presses. Source: FlintGroup.

Legal migration limits:

- 60 ppb Overall migration (OML) for substances
- Specific Migration Limit (SML) for evaluated substances
- 10 ppb for unevaluated substances

Figure 5.2 Legal migration limits

narrow-web presses, while mid-web technology is better suited to run lengths between 5,000 and 50,000.

These figures are not related to future trends, they show what is happening now. It also explains why label converters, already well-used to printing shorter runs, versions and variations, see flexible package printing as a market opportunity. Certainly, when looking at the printing/ink technologies that are used nowadays on these two market sector applications, then UV-curing, widely used by label converters, is a technology to consider.

These trends, and with them the trend to move from long runs on wide webs to shorter runs of flexible packaging printed on narrower web presses using UV-curable inks, becomes highlighted by the fact that some 70 per cent of all flexible packaging is food related – it's used to wrap food.

Therefore, the inks used to print the flexible packaging must be acceptable in terms of UV Food Contact Material (FCM) inks.

Historically, the term low migration inks was widely used. Today, there is an agreement between all the different ink manufacturers, including members of the UVFoodSafe Group, to no longer use the term 'low migration ink' but to all use 'Food Contact Material' inks. So how are these FCM inks defined?

Food Contact Material inks are inks that, if correctly applied and correctly cured, and with the right choice of packaging concept, the legal migration limits can be met. But, what are these legal migration levels?

- It's 60 PPB for the overall migration level of substances.
- It refers to SML or a specific migration limit for those, let's say, substances of which toxicological data are existing – and for those

substance that exist are considered good evaluators.
- Where toxicological data are non-existing, the limit is 10 PPB.

How can PPB (Parts Per Billion) perhaps be more simply defined or made understandable? Well, one PPB can be compared with one tea spoon of water in an Olympic-sized swimming pool. So, if we are talking about PPB, this is maybe, a much easier way for it to be visualized. Essentially, we are looking at very, very, very small quantities when we discuss migration limits (Figure 5.2).

MIGRATION

Migration is undoubtedly an important consideration when talking about flexible packaging inks and their performance, but what are the different ways that migration can occur?

The first one is **penetration migration**. What does this mean? Essentially, the ink is able to go through the substrate. If the substrate is a plastic and it's PP or PET, PVC, or whatever, it's always some form of barrier, and some plastics are better barriers than others, like PET is a better barrier, but still it is nevertheless a non-full barrier. It's a plastic, and all plastics are non-full barriers. They will allow some level of migration.

That's the first kind of migration. The second area of migration, also referred to already a couple of times in previous chapters, is **set-off migration** in the reel, whether the set-off is from outside or inside contact. This is something that needs to be taken into account.

The third way that migration can occur is via **vapour phase migration**. That is volatilisation of compounds during cooking. This will have an effect on how the ink or coating can penetrate through substrates.

The last way that migration occurs is through **condensation extraction** – that is the condensation of critical components when cooking/sterilization takes place.

These different forms of migration relating to the safety of food packaging are illustrated in Figure 5.3.

Already mentioned earlier, the three overall requirements of UV-curable FCM inks are:
Regulatory, **Performance**, **Economical**.

WHAT IS MIGRATION?

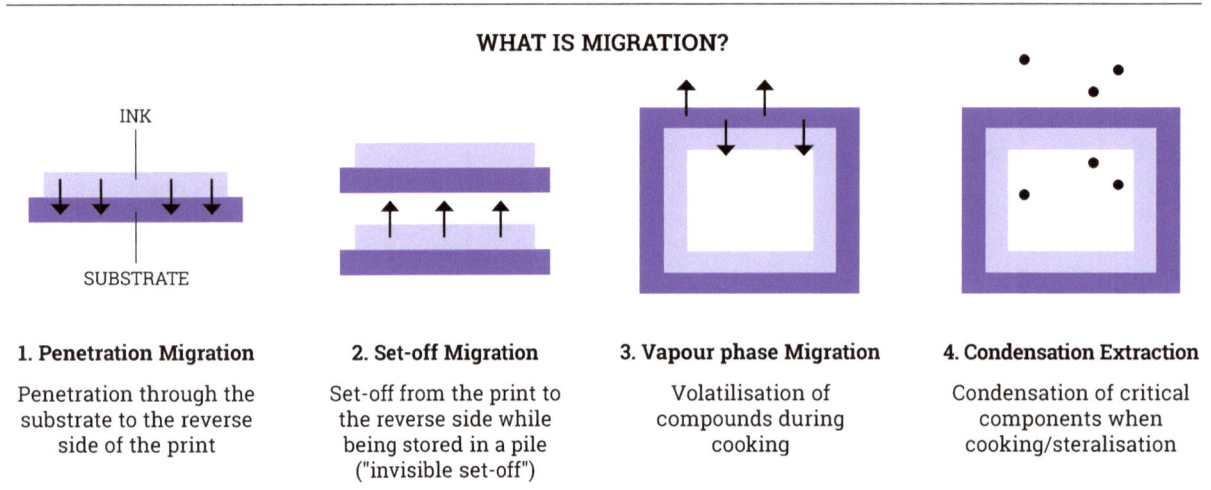

1. Penetration Migration	2. Set-off Migration	3. Vapour phase Migration	4. Condensation Extraction
Penetration through the substrate to the reverse side of the print	Set-off from the print to the reverse side while being stored in a pile ("invisible set-off")	Volatilisation of compounds during cooking	Condensation of critical components when cooking/steralisation

Figure 5.3 Shows the different forms of migration. Source: FlintGroup.

We need the **Performance** of the inks, especially after they are cured (and through all stages of production, forming, filling, distribution, and consumer usage). But, also, of course, by complying with the required **Regulatory** landscape. The regulatory landscape and the overall and various regulation and legislation requirements are driving a lot of what both the ink manufacturers and converters have to comply with today.

Overall, these regulatory requirements can be identified as follows:
- No Specific European harmonized legislation for inks but several legislative instruments which impacts materials and articles for food
- Some examples in the EU:
  - Regulation EC 1935/2004
  - Regulation EC 2023/2006: rules on GMP
  - Directive 10/2011 (Plastic Regulation)
- Swiss ordinance is the first specific and complete legislation on printing inks
- Nestlé Guidance Note on packaging inks.

The various regulations identified above oblige the whole industry, starting from the raw materials suppliers, or cell and ink manufacturer, the printer or converter, and even the brand owner, to comply with certain regulations.

The bigger challenge that the whole industry is facing now is that there is not one specific harmonized legislation for inks that exists today. But, there are several regulations that refer to them. They have just been mentioned. It's the Regulation EC 1935/2004. It's the Regulation EC 2023/2006 and there is the plastic regulation. Next to these, there is the Swiss ordinance, which was the first complete legislation on printing inks. That's why, more or less, the whole industry is always referring to the Swiss ordinance. It's legislation that is really existing.

Then, of course, there are the guidance notes from Nestlé on the brand owner requirements, and there are also a couple of other brand owners that are working in this area and have their own guidance notes – which have both a positive and a negative list. The positive list meaning ingredients that the ink maker can use; the negative list meaning things that are not allowed to be used.

It is this regulatory landscape that is pushing the whole flexible packaging industry to be compliant with these various regulations and why ink makers have to undertake several things – the most important of these in order to be compliant is, first of all, raw materials selection.

RAW MATERIALS SELECTION

Ink makers that want to make an FCM ink, need to start with selecting the right raw materials. What does

GMP: GOOD MANUFACTURING PRACTICES

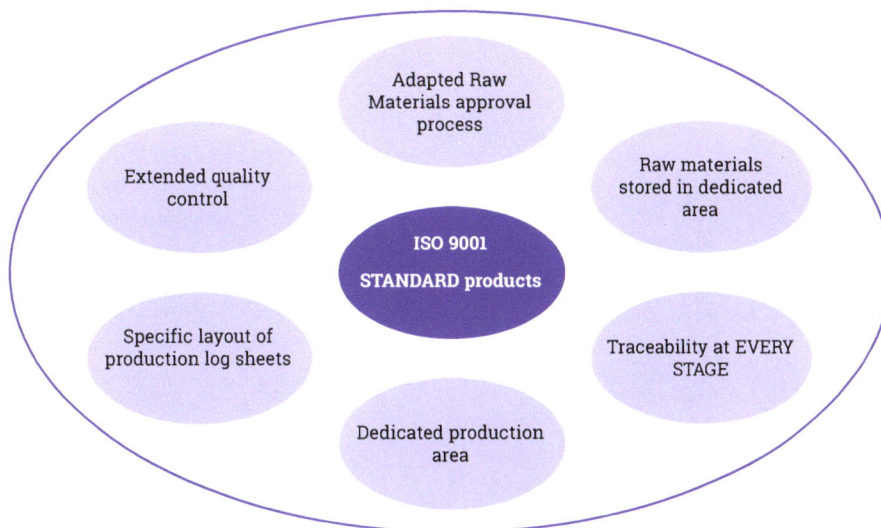

Figure 5.4 Good manufacturing processes

that mean? The raw materials should be Swiss listed and they should be compliant with all the required regulations.

Next to that, the ink maker needs to make sure that they know everything about the raw material being used. Does it have smaller molecules in there like stabilizers, like a catalyst; also any small ingredients that are there, impurities; everything needs to be known, all the constants. There are raw materials suppliers on the market today that have dedicated raw materials for this and they can be worked with. So, that's very important.

Another thing, which is driven by EUPIA, the European Printing Ink Association, is that you cannot use CMRs: carcinogenic, mutagenic or reprotoxic cat 1A and 1B raw materials. So, each and every member of EUPIA has signed this policy to not use any CMRs.

The ink industry was also confronted in 2018 with some raw materials that were commonly used in UV inks shifting from being okay to being labeled as CMR. This resulted in a lot of work having to be done in terms of reformulating all these areas.

ECONOMICAL ASPECTS
Apart from the regulatory and technical performance requirements of FCM inks, it is also necessary to consider the economical aspect of the inks as well. The flexible packaging industry ideally wants FCM inks at the same price as non-FCM ink, so they need to also fit the purpose. It's not just the regulatory and performance evaluations that, let's say, give ink manufacturers a smaller opportunity, there are also the various economical considerations that have to be taken into account. For the ink manufacturer it is quite a challenging job to put all these different and varied requirements together.

GOOD MANUFACTURING PRACTICES
It is also important for an ink manufacturer to have the right production process for flexible packaging inks. They need good manufacturing practices (GMP). These are shown in Figure 5.4. and, as can be seen, include raw materials approval, raw materials storage, traceability, dedicated production area, log sheet layout and quality control.

So, what does good manufacturing practice actually mean?

On top of the ISO-9001 standard, it is necessary to undertake a number of things in order to be GMP approved; ink makers need to have an adapted raw

Additional considerations	Design window: Standard UV ink
• Less building blocks available • Excluding low molecular weight monomers prone to migrate • High colour strength • Legislations and regulatory environment continuously moving • Increased awareness among local brand owners • Fit for purpose	Available number of ingredients available to formulate UV inks in general

Design window: Food packaging compliant inks

Smaller number of substances available, with FCM window continuing to shrink

The above are additional considerations a Formulator for FCM inks needs to take into account

Targeting a design window that gets smaller every year

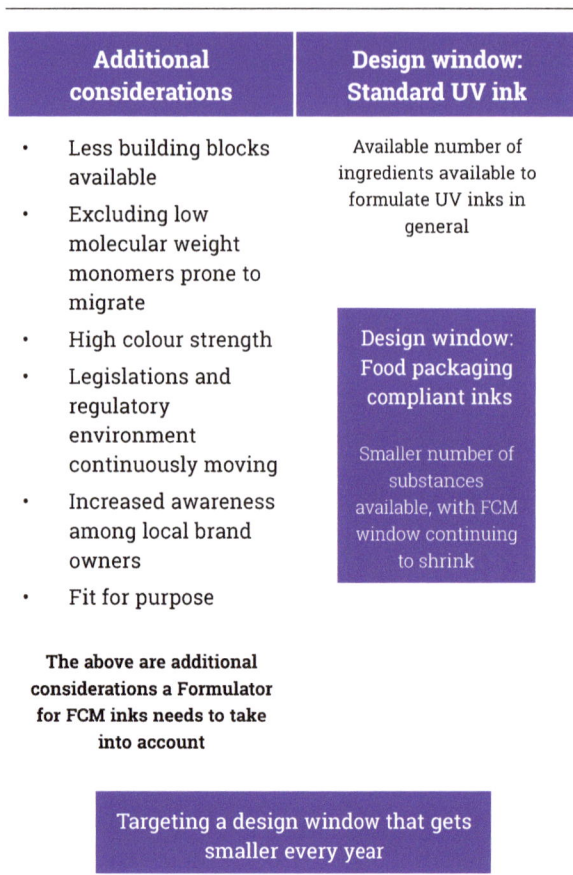

Figure 5.5 Raw materials: designing FCM inks

materials approval process. All the raw materials being used need to be the right ones. They need to be stored in a dedicated area.

The production area also needs to be dedicated. For instance, at Flint there are two floors; the second floor it's only for FCM inks. The FCM inks are only made on that floor. Everything is completely separated.

There needs to be traceability, very important, at every stage, so that if there is any kind of problem each and every company in the value chain or in the supply chain is able to track back so that the problem can be identified and rectified. Traceability is therefore very, very important in the manufacture of FCM inks.

Next comes extended quality control that

manufacturers also need to organize, record and document the GMP, which is not only value for the ink manufacturer, but also for people down the value chain. That's why recommendations that we make to printers and converters need to be followed.

These recommendations are preferably to have dedicated presses to ease the handling, and avoid contamination of the press. If not possible, very rigerous cleaning processes are required in between different print jobs. So, when a job which is a non-food, or a job which is indirect food, is done on the same press; clean it very, very well. It's extremely important. If there are special products being produced the ink supplier can also provide guidance on the process – but, again, dedicated presses are the ideal case.

DESIGNING A UV FCM INK

Coming back to the raw materials and the regulations and the limitations that the regulations provide, as these get more and more stringent, the ink maker is targeting and designing for windows that get smaller every year. Designing a UV, FCM ink, is now within a design window that has shrunk considerably because of the regulations. These mean no CMR and so forth. There are fewer building blocks available to make the ink (see left hand side of Figure 5.5). Yet, still, the technical performance should be the same as non-FCM inks, with regards to colour development, with regards to adhesion, printing speed and so.

INK CURING IS CRITICAL

The critical parameter in the whole UV, FCM ink process is the curing aspect. Everybody in the value chain, until the converter, can do a fantastic job in having good and proper and non-CMR raw materials, put them together in an ink – and then not process it following a proper Good Manufacturing Process (GMP) at the converter. The final, very important step in the process, is the curing. Other things are also important, but if the ink is not properly cured the chances of having migration occur are higher than when it's properly cured. So, why is curing so critical?

First of all, there is a higher risk of migration and this is demonstrated by facts. There is a consortium in the industry, called the UVFoodSafe Group, which includes ink manufacturer, materials suppliers, and press manufacturers. This Group undertook a trial

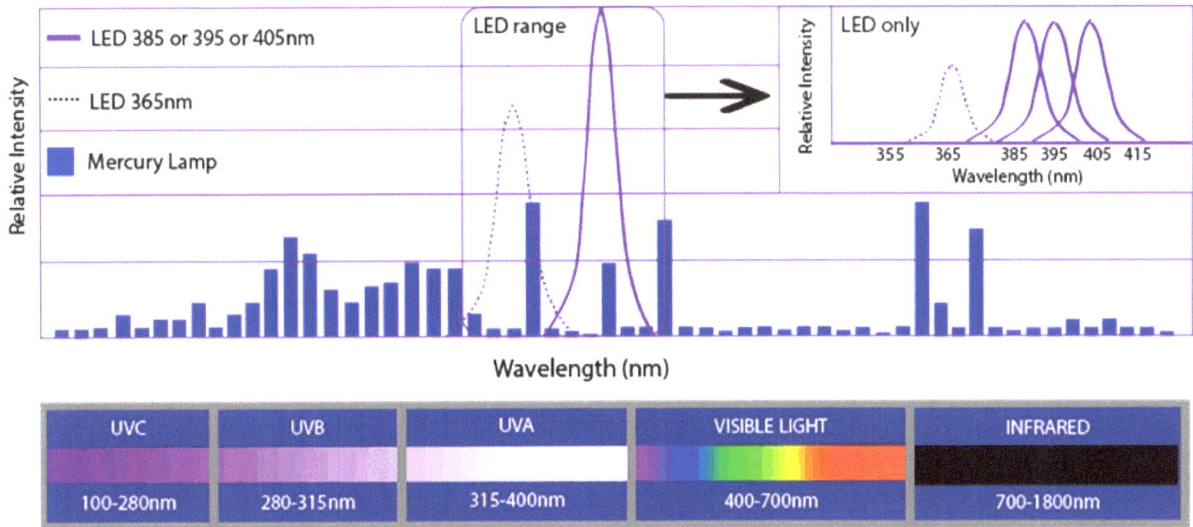

Figure 5.6 Wavelengths of UV Mercury and UV LED lamps. Source: Phoseon.

together which demonstrated that the less curing that takes place, the higher the migration.

When people say they don't want migration, it should be understood that, at science levels, there is always going to be some migration. It's better to control that migration and to know what is migrating because, in practice, there is always a little bit taking place. It's better to know which specific substance is migrating, to know what the limit is, to control this. It's much more important than to say the aim is zero migration.

Without proper curing in FCM applications there will be less good scratch and solvent resistance.

So, what do printers/converters need to do to ensure proper curing?

- **Monitor the printing speed**. The higher the speed is, the higher the risk of having higher migration. Each and every aspect needs to be taken into account. Next to the curing there is also the thickness of the ink. There is the substrate, there is the adhesion. They have multiple parameters, but the UV curing speed is important.
- **Monitor chill roller temperatures**. If you have a very thin substrate, heat sensitive, it needs to be cooled. Fully understandable is that heat

helps the curing process. So, it is necessary to find the right balance between how far to cool or heat the substrate up a little bit – so that the substrate is not damaged.
- **Monitor the ink thickness**. The thicker the ink, the more difficult it is to cure. So, that's another aspect that can be monitored.
- **Check the UV lamp intensity** and check the lifetime of the UV lamps because intensity of a UV lamp is going to go down in time. That is also something that needs to be monitored.
- **Check UV lamp reflectors**. Make sure they are clean and doing their job effectively.

Now, let's consider the sources that produce UV light. There are two UV light sources nowadays that are predominately used on the market. One that has already been there for many, many years, which is the UV Mercury lamp. It's very well known by the label printers. Then, more recently in the last couple of years, there has also been a big increase in the use of UV LED lamps.

As can be seen in Figure 5.6., UV light Mercury has an output wavelength between 200 and 400 nanometers, while UV LED is only at 385 and 395 nanometers.. It is monochromatic light that is shining on the ink. The wavelength and the light intensity dictate the chemistry.

The UV Mercury is referred to as conventional UV - and it should be noted that conventional UV inks will not cure when exposed to UV LED light.

So, from an ink perspective, if label or flexible packaging printing customers go to UV LED, there is a different chemistry that is used. This means that at Flint for example, they have FCM inks for UV Mercury and they have also UV LED FCM inks that are available.

It should also be noted that there are different UV LED lamp suppliers on the market. Some of these use 385 nanometers, others are 395 nanometers. It is therefore important for the printer/converter to know and understand all the parameters that could influence the curing. These include:

- The irradiance output
- The distance between the ink and the lamp
- The angle of the lamp to the web.

As previously stated, it is also important to know the press speeds and so forth. All the parameters that could influence the curing need to be checked, with

Figure 5.7 A pane of glass is a functional barrier. Paper is not

job printing conditions documented very well.

BARRIERS, BARRIER COATINGS AND LAMINATED WEBS

Chapter 2 of this book looked in quite some detail at

OVERALL REQUIREMENTS: REGULATORY: MIGRATION TESTING SET-UP

Determination of the amount of migration from cured samples through a migration cell

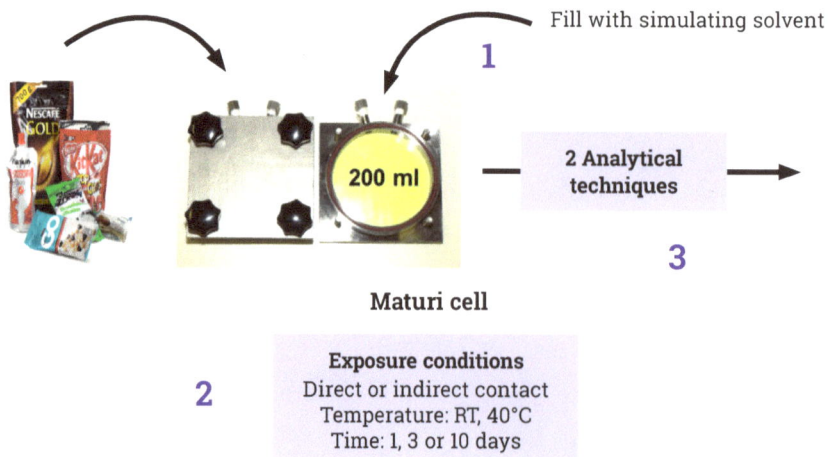

Fill with simulating solvent

1

200 ml

2 Analytical techniques

3

Maturi cell

Exposure conditions
Direct or indirect contact
Temperature: RT, 40°C
Time: 1, 3 or 10 days

2

Figure 5.8 Migration Testing Set-up

FOOD TYPES	FOOD STIMULANTS
Aqueous, non-acidic and non-alcoholic foods	10% Ethanol (in water)
Alcoholic foods (<20%)	20% Ethanol (in water)
Dairy products and high-alcoholic beverages (>20%)	50% Ethanol (in water)
Acidic beverages (pH <4.5)	3% Acetic acid (in water)
Fatty foods	Vegetable oil (substitute 95% Ethanol in water)
Dry / high temperature food	Tenax = poly (2, 6-diphenyl-p-phenylene oxide) 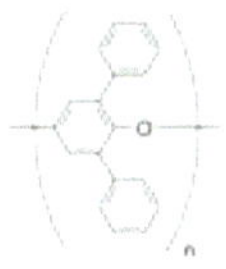
UNIVERSAL STIMULANT	**95% ETHANOL, 5% WATER**

Figure 5.9 The choice of food simulant in migration testing.

the substrates used for flexible packaging, and stating that there is no perfect material, which means that many materials used for flexible packaging involve the use of multi-layers which are created by bonding together (laminating) two or more materials, whether paper, plastic or foil, to create the required barrier.

Mention was also made in Chapter 4 of barrier coatings that can be applied to flexible packaging to prevent migration of ink, adhesive, or other substance through the face material. Barrier coatings provide barriers for food packaging requirements, and may include protection against oxygen and aromas, liquid water and water vapor, oils, and grease. An effective barrier can prevent both losses from the packaged product, and penetration into the package, both of which can affect quality, and shorten product shelf life.

Packaged food products are being maintained fresh longer as a result of new materials, and food processing developments. For example, O^2 scavengers are now being used that work within a sealed package to limit O^2 reaction with a food product. Combined with effective O^2 barrier packaging, food packagers have the ability to improve shelf life, preserve product appearance, and flavor, while minimizing preservative use.

Knowledge around migration and the various methods of barrier protection is therefore crucial for servicing the flexible packaging food markets. Suppliers must be able to provide the right barrier or

---

Overall requirements: Regulatory: Choice of the exposure test conditions

- Indirect contact with food simulant
- Temperature: RT, 40°C, ...
- Duration: 1, 3, 10 day, ...

Migration testing must be conducted under the most severe conditions of temperature and time anticipated for the proposed use.

FOOD APPLICATIONS	TEST TEMPERATURES	TIME
Room temperature	40°C	10 days
Refrigerated or frozen	20°C	10 days
Microwave and oven	120°C	30 min to 1hr

Figure 5.10 Choice of exposure test conditions.

OVERALL REQUIREMENTS: REGULATORY: IDENTIFICATION AND QUANTIFICATION OF MIGRANTS

Liquid Chromatography with Mass Spectrometry (LC-MS)	Gas Chromatography with a Mass Spectrometry detector (GC-MS)
• Both volatile and non-volatile compounds • Identification of migrants in food simulant • Quantification of **low and High Mw ingredients** (Mw < 2500 Daltons)	• Volatile and many organic-soluble analytes • Identification of migrants in food simulant • Quantification of **low Mw ingredients** (Mw < 500 Daltons)

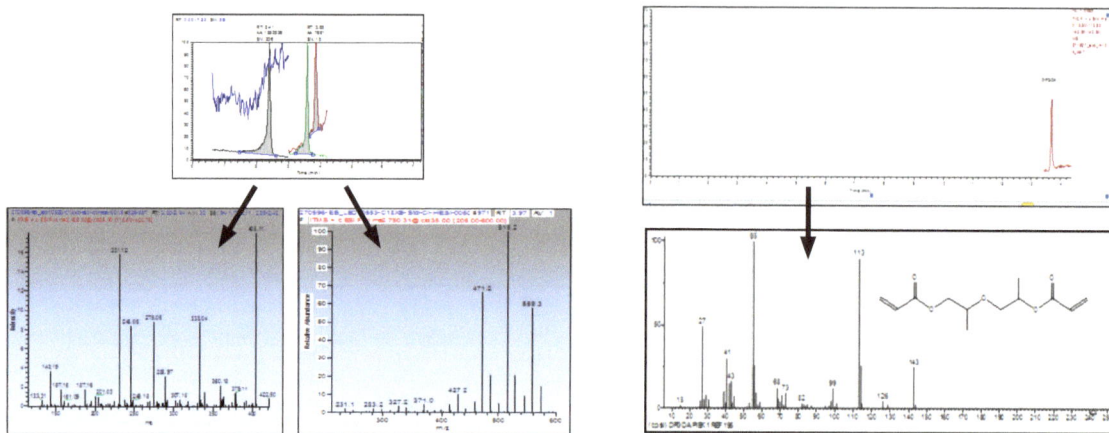

Figure 5.11 Identification and quantification of migrants

seal for a given application and be able to support printers and converters with 'fitness for use' testing.

But, just to be clear, what do we actually mean when we talk about a barrier?

Let's use Figure 5.7. to explain. In this, we can say that a pane of glass is a barrier. Try putting your face to a pane of glass at home. You won't get through it. It's an effective barrier. Aluminium of a certain thickness of say, 12 micron and above, is also an effective barrier.

What is not a barrier? Well, paper and plastics films. Even if you have a PET or polypropylene films from 40 or 38 micron, they are not an effective barrier. There are always voids in film. It is never 100% closed. Migration is able to occur. The thicker the plastic, the longer it will be before migration takes place – but it will still occur.

Providing barrier layers, whether through laminate

constructions or the use of barrier coating, are properly constructed and applied, and everything is done as it should be, there should be absolutely nothing to worry about.

MIGRATION TESTING

To build further on migration and all the related regulatory, production, performance and curing requirements, how is migration testing carried out? The overall requirements are illustrated in Figure 5.8.

Migration testing is the process in which we are measuring the amount of migration from a cured sample into a migration cell: the printed package is brought into contact with a simulating solvent. It's not the ink itself that is in contact, it's the other side of the print. The cell can also bring in different temperatures. It can even be kept for different types of printed packaging at specific temperatures. The third part is

COLLABORATION THROUGHOUT THE WHOLE VALUE CHAIN

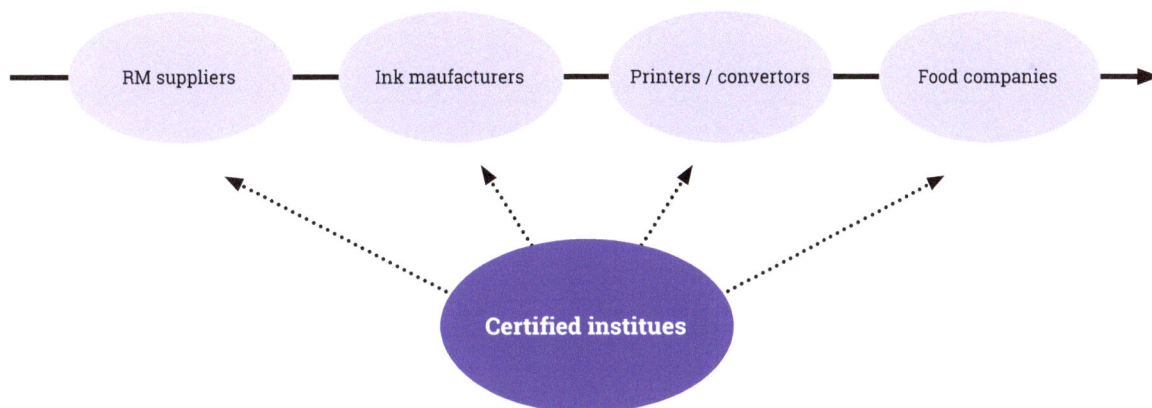

Figure 5.12 Collaboration throughout the whole value chain.

to use the right analytical techniques.

Depending on the food type that is going to be wrapped and the plastic that Is being printed, and depending on the food type being packaged, the right simulant needs to be chosen. This can be examined further by studying Figure 5.9.

As can be seen, for fatty foods the simulant needs to be a vegetable oil (in water): for dairy products it is 50 per cent ethanol (in water), and so forth. Because of this wide variety, the industry accepted to have one a universal simulant, which is 95 percent ethanol (5% water).

The next element of the test is how long should the non-printed side of the sample be in contact with this simulant in the cell? How is the choice of exposure test conditions determined? This depends on the way the food is going to be used. Will it be used for example, at room temperature, in the refrigerator or in the microwave? The answer will have an impact on the temperature and the time that is needed need to bring the non-printed side of the sample into contact within the cell. This is shown in Figure 5.10.

The most common condition used is 40 degrees Celsius for ten days. This is the generally accepted and most widely used whole industry test method to do migration testing. Bring the non-printed side of the

sample in contact with 95 percent of ethanol for 10 days at 40 degrees Celsius.

After 10 days, analytical techniques are used to examine what's inside the cell. The institutes that are performing migration tests will therefore need to know what to look for; a statement of the composition of the ink is always given to the institute doing the testing. They will use two techniques

- Liquid Chromatography with Mass Spectrometer (LCMS) and Gas Chromatography with a Mass Spectrometer detector (GC-MS)
- to look for possible migrants that are coming from the ink. Figure 5.11.amplifies the way that identification and quantification of migrants is undertaken

CONCLUSION

Within the flexible packaging industry, both in Europe and outside of Europe, there are printers and converters using UV Mercury and UV LED FCM inks. They are safe and working to produce UV printed flexible packaging, and there are even big brand owners and small brand owners that are already using and accepting this technology for indirect food containers.

The same applies to shrink sleeves. Do not forget that the shrink sleeves are printed on the inside of the sleeve. However, if the bottle is a plastic, ink migration can take place through the bottle. Yes, the bottle is normally thicker and the chance of migration is lower, but in principle, scientifically speaking, microscopic levels can still be there.

Nevertheless it should be emphasised that UV curing is a demonstrated technology that's been working for flexible packaging. To continue this technology's success, it is very important that there is an open collaborative spirit throughout the whole supply or failure value chain. See Figure 5.12.

It's important that everyone in the supply chain works with each other in an open way, starting with oversight of the supplies.

Supplies of the raw materials to the ink manufacturers are very important. Then, the ink manufacturer in turn, needs to be giving full information to the printers and the converters and on the food companies, as well as full information to the institutes that do the migration testing. All the different possible migrants and all the different ingredients in the ink need to be given to the institutes so that when they receive a sample from a printer to do migration testing, they know exactly what to look for.

It is also recommended that the printer/converter also implements GMP if possible – dedicated presses, monitoring of the ink, the ink thickness, the temperature of the chill roll, the substrate being used. The UV curing; whether this is LED or Mercury, and also very important that they know what the printed packaging is going to be used for. What is happening after printing? Will the pack be going through sterilization; is there sealing and so forth?

The message is to only do the things in the right way, promote UV technology to the brand owners, and to perceive that it is a safe technology when undertaken correctly.

Chapter 6

Understanding forming, filling, sealing and lidding operations

When labels have been printed and converted into their finished size and shape they will be sent in reel or cut-stack format to the packaging plant for applying to the required bottle, can, jar, tub or other packaged product. Flexible packaging goes through a similar type of printing and converting process before being sent to the packaging plant in slit reel or die-cut format to either go through a forming, filling and sealing machine operation, be flow-wrapped around a bar or other solid product, or applied as lidding or capping to filled pots, tubs, jars, etc.

In Chapter 3 the many different types of flexible packaging products were described. They included wrappings, bags, sachets, pouches, and lidding. Some packs were noted as being quite small, or long and thin, while others can be fairly sizeable. Some need to be filled while the pack is in a vertical position, using either a single web or a dual web machine; others need to be filled or wrapped horizontally.

Whatever the size and shape of the pack, a means of either pre-measuring or weighing the product or product dose going into, say, a vertically filled pack is required, or the already bottom sealed pack is lowered onto a precision weighing table and the product to be packed is dispensed until the gross weight of the product filled pack has been reached, at which time the filling stops and the top is sealed (which also forms the bottom seal of the next pack) and the pack is cut from the web. It will then go on for boxing and shipping.

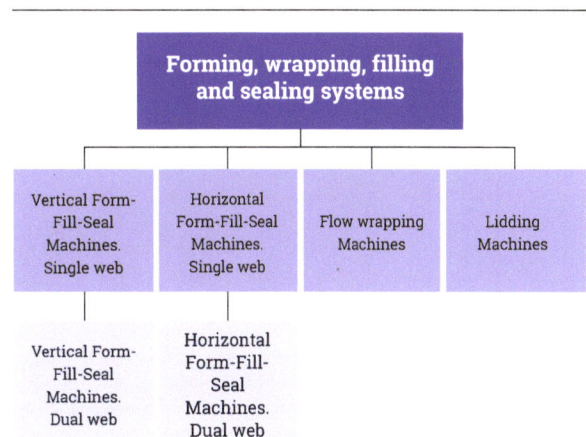

Figure 6.1 The main types of forming, wrapping, filling and sealing machines

There are different sealing methods available (heat, adhesive, crimping) depending on the packaging material being used and the application. Such a range of requirements, in turn, means there are many different types of machines used to form, fill and seal the variety of packs, to wrap products, or to apply lidding (Figure 6.1).

The nature of the product to go into the various types of flexible packaging will also have an impact on how the printed reel or lid is formed, filled and the pack sealed. Is it a liquid, a powder, granules, cream, paste, a solid, soft or hard? How does that impact on the machine design and operation? What kind of transporting system, hopper, weighing or dispensing arrangement is required to bring the product to the filling machine and each individual pack?

To reduce crushing of the product in transit and use, extend shelf life, or perhaps retard the growth of bacteria, the final sealing process may require the bag, sachet or pouch to also be filled or inflated with air from a blower or from an inert gas supply such as nitrogen – which is a natural means of extending shelf life.

In addition, there may be a requirement to provide tear notches; to cut out hanging holes or slots; to provide carry handles, to add a recloseable feature; to provide some kind of spout or pouring device. For some applications the filled packs are not separated, but rather perforated and supplied in strips for individual packs (sachets) to then be removed and used progressively, such as from a hanging display.

The packaging machines that form, fill, close and seal a flexible package in one continuous or intermittent-motion operation can be configured so that the printed web travels horizontally through the machine (horizontal form-fill-seal) or vertically through the machine (vertical form-fill seal).

Fed from a roll, the web is folded to the desired package shape and stabilized by heat sealing. The product being packed is placed into the formed package, and the remaining opening is sealed. There are many different types of machines that perform these operations.

Rather than trying to describe and discuss all the many different types of forming, filling, sealing, wrapping or lidding machines, this Chapter simply aims to explain the principle of the main types of equipment on the market. These can perhaps best be summarized as:

VERTICAL FORMING, FILLING AND SEALING

Vertical forming, filling, and sealing machines are fairly complex, yet flexible and versatile machines used in the packaging industry to package various types of free flowing food, oils, powders, other materials and a wide range of liquid products.

Speed of packaging, ease and convenience of

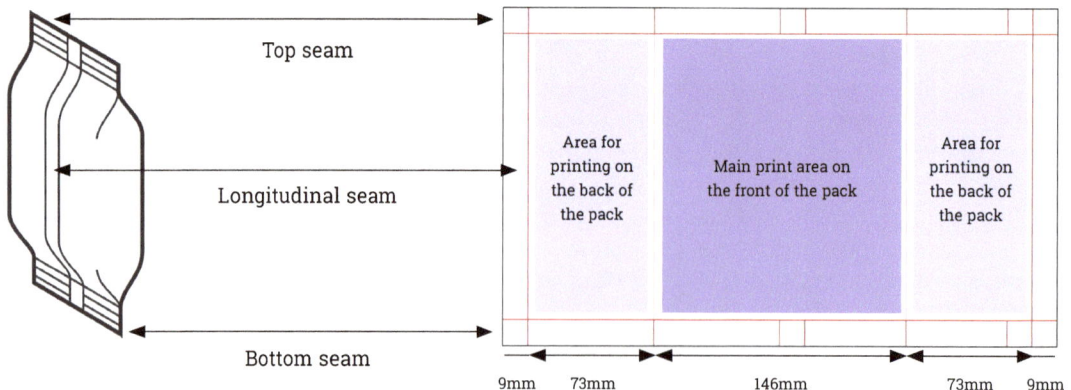

Figure 6.2 How the flat printed substrate ends up as a sealed and seamed pack. The areas available for printing are also indicated

changeovers, machine reliability, low maintenance, and low operational costs have all helped to advance their popularity of this type of equipment. Other important factors include software oriented operation, increasing demand from the food and beverage sectors, chemicals and personal care industries, and growing demand for convenient packaged consumer products.

These types of machines enable the packer to have a vertically operated packaging system in which:

- The printed flexible packaging material is formed into the required sachet, bag, tube or pouch shape, the two outer edges of the pack being joined together with a longitudinal or vertical seam
- The bottom edge of the pack is also then sealed
- The weight or volume of product being packed is measured
- The already formed shape is filled with the product
- The top of the newly filled package is sealed.

This can perhaps best be illustrated in Figure 6.2.

Vertical forming and filling means there is no need to have a large amount of floor space available to place the machine when compared with horizontal machinery. They are therefore quite compact.

Packs made on this type of equipment are either normally flat, or made to incorporate gussets at the sides, or provide block (flat) bottoms to enable the pack to 'stand up' on the retail shelf or in the consumer's cupboard or fridge. Both types have a large flat front and back surface that will enable printing on both sides of the pack (albeit that the back of the pack will usually have a vertical center seam area that is not printed on), making the most of the available space. Two key types of vertical form, fill and seal machines are commonly used:

- Single web machines
- Dual web machines.

Single Web Machines. The manufacture of a vertically formed-filled and sealed pack is illustrated in Figure 6.3 and starts with the flat printed web approaching the back of a long hollow conical tube. This is known as the forming tube. As the center of the web approaches the tube, the outer edges of the substrate form flaps that wrap around the forming tube. At the same time, the web is pulled downwards (this is where the correct coefficient of friction/slip is

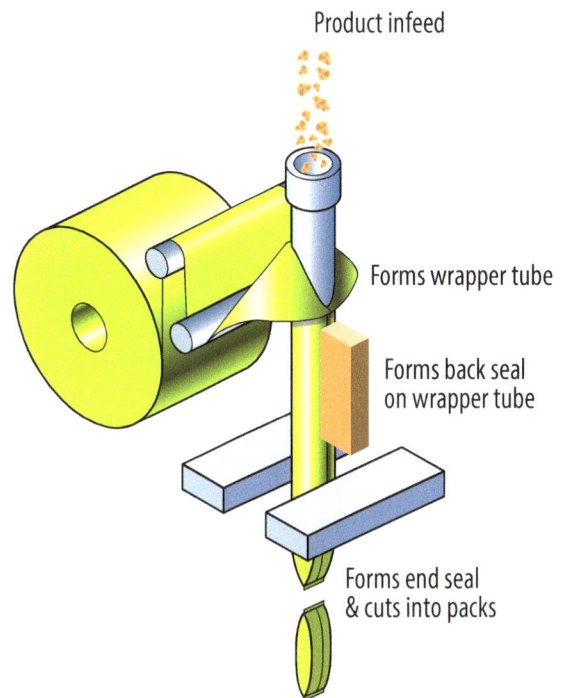

Figure 6.3 Diagram shows the stages in the operation of a single web form fill and seal machine

important (see Chapter 4) around the outside of the tube and a longitudinal/vertical sealing jaw clamps onto the edges of the web to create a 'Fin Seal', bonding the edges together by melting the seam edges together or by using an adhesive. Ultrasonic sealing is a new development which tends to be used for heat-sensitive products and permits sealing through liquids.

At the beginning of the bag filling process, a horizontal cross sealing jaw creates the 'bottom seal' by clamping and sealing across the bottom edge of the now tubular bag. This bonds the bottom edges together, and cuts off the previous filled bag and/or any bag material below the seal.

The feeding of the web of material and cutting of the bag/pouch depth to the correct size can be determined either by pouch length, or by indexing to a photo registration mark, which is then detected by a visual sensor.

The product being bagged is now dispensed

through the long conical filling tube into the center of the bag. When the right content quantity of the product-filled bag has been reached the filling stops, and the horizontal cross sealing jaw seals the top of the bag. Any required product finishes, such as hole punching for retail hanging racks, will be done concurrently or just after this 'top seal' is made.

At the same time this also simultaneously forms the bottom of the next bag being formed above. The filled and sealed bag can then be cut off from the tube to provide an individually sealed package, which is then free to move onwards into the required filled product packing and shipping operations.

It should be noted that each bag, pouch or sachet material, as well as the specific end-use application, has a bearing on the heat, pressure and dwell time sealing jaw settings necessary for creating the filled pack. The preciseness in which these three factors are controlled and balanced will determine to a great extent the quality of the seal.

The amount of time the seal jaws are closed and in contact with the sealing material is generally defined as the 'dwell time'. This is adjustable on most forming, filling and sealing machines and is a tool to ensure the correct temperature is obtained on the sealant film.

The pressure applied between the sealing jaws when the heated surfaces are in contact with the material is also a critical adjustment, ensuring that the material is held in place without wrinkling until the heat is driven into the seal area. Excessive heat/temperature can create as much problem in producing a continuous heat seal as too little heat If sealing issues arise it is important that the heat, pressure and dwell time characteristics are adjusted during trouble shooting to uncover the root causes. A change to any one of these factors will mean that one or both of the other two must also be adjusted to compensate for the change.

A single lane machine format is used for various sizes of pouches or bags as it forms a single pouch during each cycle. Multi-lane format machines create multiple pouches in each cycle. They are used for multiple small packets such as small salt and sugar packets.

Recent advances in bag, pouch and sachet forming machines and systems technology have enabled even smaller and smaller vertical form, fill and

seal machines to be developed. In particular, multi-lane sachet machines are now being used in a multitude of industries for small quantities of powder and liquid products. Sealed on all four sides, these small packets are often filled with condiments and spices like salt, sugar, mayonnaise, or ketchup.

Manufacturers and contract packagers love sachet machines because they can achieve high throughput while occupying limited floor space. The operation of multi-lane sachet machines is similar to that of other form, fill and seal machines, but with a few notable differences. Basically, a large roll of film is unwound by motor driven rollers and kept in constant tension. It is then slit into two halves which are brought together and formed into multiple sachet packets, filled with the product, and then sealed, all in a vertical fashion, at rates of up to 80 cycles per minute, per lane.

A group of pneumatically-driven disc knives can then cut the film vertically if individual sachets are desired. A horizontal cutting station then separates the filled, sealed, and longitudinally cut packets into individual sachets. A single sachet machine with the capacity for up to 10 lanes can produce up to 800 packets per minute, or about 13 per second. However, the more lanes there are on a multi-lane sachet machine, the fewer opportunities there are for narrower web converters.

Dual web machines. Dual web form, fill and seal machines are used for the manufacture of four side sealed bags or pouches, as well as for packs that require a different materials to be used on each side. Dual web systems use two rolls of material instead of one, which are fed in from opposite sides of the machine. The bottom and sides are heat sealed together to form the pack, and the product is again loaded from the top. The pack with the loaded product then advances downwards; the top is sealed and the pack is cut off. As with single web machines, the sealing of the top of the pouch forms the bottom of the next pouch. During this process a tear notch or a hanging slot may be added.

Whether single web or dual web, vertical forming, filling and sealing machine may be divided into single, multiple lanes and open and closed machines.

In terms of the end-use industries and applications vertical forming, filling and sealing machine are used in, these include:

• Food and beverage industries for tea, sugar,

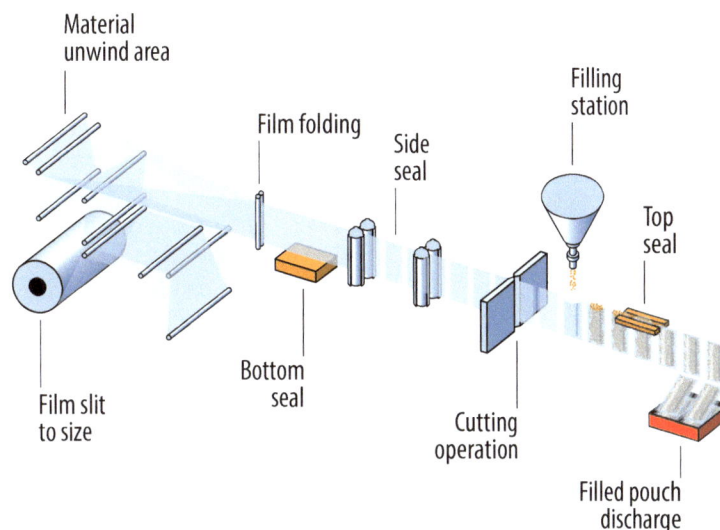

Figure 6.4 Operation of a typical horizontal pouch forming, filling and sealing machine

wafers, and candies
- For fresh vegetable packaging and pouches for sauces and detergents
- For chemical and pharmaceutical industries for various packaging purposes
- The packaging of bulk goods such as nuts and cookies
- For powdered goods such as coffee, for grains and granules such as detergents
- For tea and liquids such as ketchup, and gel.

HORIZONTAL FORMING, FILLING AND SEALING

In addition to vertical form, fill and seal machines, there are also horizontal packaging machines that work in roughly the same way. Closely related in terms of overall operation, horizontal form-fill-seal machines will generally use more floor space than a vertical system, with all the stages of unwinding, slitting, folding, sealing, cutting and filling carried out in one horizontal inline operation. This can be seen in Figure 6.4.

As can be seen, the horizontal form, fill seal machine has a reel unwinder on which the web is mounted. As the web is released from the unwinder it passes through a forming plough and is then sealed

at the bottom and the two sides. A cutter will cut the pouch vertically at the right time, being controlled by a photocell. An opening device then opens the pack and brings it to the filling station. The top stretching mechanism stretches the pouch or pack to make a clean top seal before the packs or pouches are discharged. It is an intermittent system in which the pack is sealed on three sides and then later filled and sealed on the fourth side.

Some horizontal form, fill and seal packaging solutions include thermoformers that produce a variety of packaging types. These types of horizontal form fill seal machines produce flexible to semi-rigid and rigid packages capable of vacuum, modified atmosphere packaging (MAP) and much more.

Many food filled packages are filled with nitrogen to extend shelf life. Food manufacturers are often looking for ways to improve their geographical reach or otherwise extending the shelf life of their product without the use of chemicals. Nitrogen filling is a natural means of extending shelf life. More and more manufacturers are choosing to create and control their own nitrogen supply by using on demand nitrogen generators.

Stand up pouches are also made on horizontal

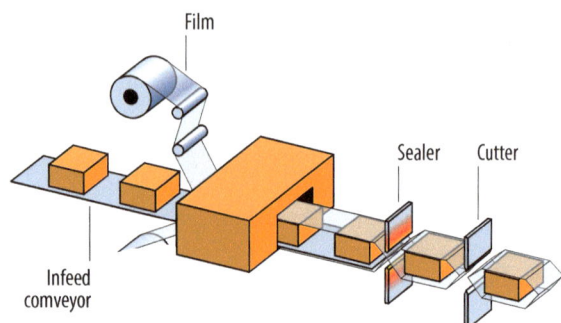

Figure 6.5 The basic principle of a horizontal flow wrapping machine

form, fill and seal machines using a film or a laminate, giving the packaging an aesthetic appeal. Other features can be added to stand up pouches, such as a spout, zip or a straw. Spouted pouches are either top spout or edge spout depending on the requirement.

Spouted pouches are used for products which are commonly free flowing liquids and pastes. Zipper pouches are used when the pouch needs to be reclosed after a required amount of product has been used so that the product remains fresh for a longer duration or retains certain properties like aroma in coffee. Straw pack is also another application wherein the straw is already made available within the pouch. The most common application for this is in beverage pouches.

FLOW WRAPPING MACHINES

Pre-printed wrappers and decorative bands applied on flow wrapping machines are an effective way of packaging and branding products. Such wrappings are typically used on confectionery, sweets, gift wrapping, butter packs, etc. In many cases products are completely over wrapped in paper, foil or film to form an airtight seal.

Flow wrapping machines provide a close fitting and air tight bag or pack around the product being packaged. They are horizontally operated, with the packaging material mounted above the operating

level. Typically, the product is loaded horizontally with a longitudinal seal formed below the pack. This can be seen in Figure 6.5

Every flow wrapping machine has an infeed conveyor, a film feed assembly (backstand), a film forming area (former), a centre and bottom seal (finseal), a cutting head and a discharge area. Food and non-food products packed by this type of flow wrapping equipment are placed on the infeed conveyor. This can be done by hand feeding or by using a semi- or fully-automated feeding solution.

An infeed conveyor delivers product to the forming area, where film is drawn from the film feed assembly into the forming area. A tube is then formed around the product and a finseal created. The film tube, now containing the product, is then delivered to the cutting head. The cutting head creates the end seals while it cuts apart adjoining wrapped products into individual packages. These are delivered to the discharge area, from where they can be either cartoned at a packing station or accumulated for packing at a later time.

Both flow wrap and over wrap methods typically use polypropylene film (BOPP) to create the wrap, although other film, foil and paper substrates can also be used.

Cold sealing methods are used, particularly in confectionery applications. The adhesive is coated onto the material. During the wrapping process the material is folded onto itself and sealed via a sealing wheel.

LIDDING

The range and variety of flexible lidding applications and machinery is quite large and continues to expand. Flexible lidding today, is used for packaging an enormous variety of foods, as well as medical items, pharmaceuticals, health, beauty and personal care products, hardware, small electronic parts and much more, in containers such as tubs, pots, trays or jars. Typical lidding materials include paper, metalized PET or PP. Quite widespread is the use of aluminum foil for lidding, particularly for water cups, dry foods, yogurt pots, ice-cream, etc. A typical lidding film construction is illustrated in Figure 6.6.

Choosing a lidding film can be a complex process as many things need to be considered: the type of seal, the seal strength, oxygen- and moisture-barrier requirements, and anti-fog requirements for the application. There's also film thickness, the ability to

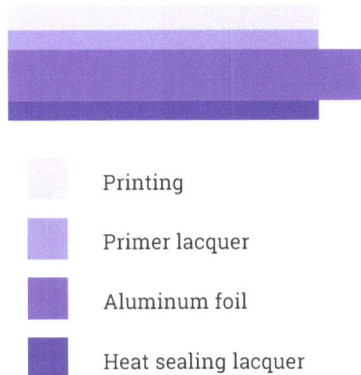

Printing

Primer lacquer

Aluminum foil

Heat sealing lacquer

Figure 6.6 Structure of a typical aluminum foil lid

Figure 6.7 A heatsealable polyester film which peels cleanly from trays in ambient or chilled conditions. Source: KPeel 3G - KM Packaging Services

seal through overfill, the type of container, the backing substrate, the ingredients inside the packaging, shelf life, sealing equipment, compliance requirements and the application conditions. All of these elements must be understood fully in order to ensure that the proper lidding material and sealing method are selected for a successful application. Quite simply, the packaging converter needs to sit with client to review all the lidding requirements, the substrates to be used, laminate constructions, sealing methods and legal requirements, and from this undertake any necessary due diligence on materials and production.

Lidding films not only seal and protect the product but can perform an important decorative, marketing, legislative and informational function. Most lidding materials are designed to be peelable to allow easy access to pack contents (Figure 6.7). Peelable seal lids require a polymer layer on the inside to facilitate the heat sealing.

Some lidding applications require a heat seal coating applied to the film construction. The coated film passes over a pre-heat station where it is warmed before it is sealed to the tray or pack, via a sealing bar or platen set at a desired heat, pressure and dwell time. The sealed tray or pack is then die-cut to shape. Lids with peelable seals can be produced to adhere to almost any material that a pot, tub or tray might be formed from. Pot applications such as yogurts use

heat seal pre-cut lids or diaphrams matched to the shape of the container.

Adherence of lidding materials to a specific pot, tub or tray and the resultant seal integrity requires a solid understanding and expertise in both film, tray and fill/seal assembly. Commonly, the lidding film will be reverse surface printed before lamination. Consideration must also be given to how the product is loaded and fed into the pots, tubs or trays in the machine, as well as how you the filled packs are handle the once sealed.

In a book of this type it is not proposed to go into all the different types of filling and lidding machinery. Some machines use in-line thermoforming to form a tub or tray, fill this with the product being packed and then apply and seal the lidding materials (see Figure 6.8).

Rather than forming the tubs or trays in-line on the lidding machine, other types of lidding machines may start with nested ready pre-formed or molded pots or tubs. The machine will then start with a denesting process, placing the individual tubs on an in-feed for filling, then applying lidding from a continuous web or stack of die-cut lids, and once again sealing and cutting.

If required, the complete line may also apply date, time, use-by or other overprint requirements. If Modified Atmosphere Packaging (MAP) is also required, then the machine will need to be fitted with a vacuum and gas flushing system. These systems allow the air to be replaced with a gas mixture during the seal process.

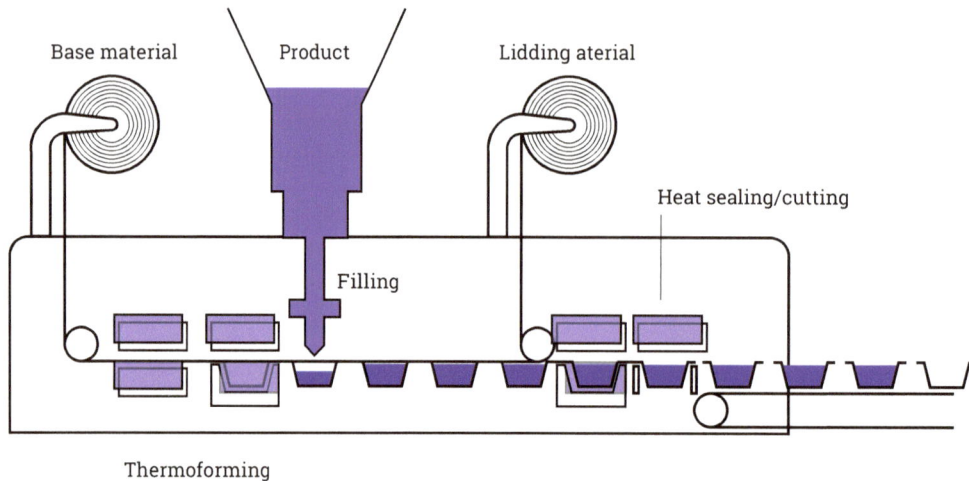

Figure 6.8 Thermoforming, filling, lid application, sealing and cutting

Normally MAP capability is defined at the outset when purchasing a machine, but most machinery suppliers can retrofit this capability if needed.

Sealing of lids is undertaken using heat sealing. Heat seal and lidding adhesives for packaging can be supplied from a number of chemical companies with very different attributes. To select the best adhesive for a specific application, it is important to consider both the end use and the converting machine process that makes the final seal.

The specific type of adhesive selected for a project is important because the substrate to which the adhesive is applied is most often different than the substrate with which it needs to bond.

Understanding the filling and sealing equipment is also critical. In addition to the proper activation temperature for a given adhesive, consideration needs to be given to the amount of pressure being applied at the temperature required to complete the seal, and the dwell time (the duration of time that the heat will be applied to the adhesive) – both of which are factors determined by the equipment and processes. Both will have an impact on the types of adhesives recommend.

Chapter 7

Markets, applications and opportunities

As earlier outlined in Chapter 1, flexible packaging has one of the highest growth rates across all printing sectors, achieving an annual global growth of close to five percent – even higher in the Asia Pacific and Indian regions. Europe and North American growth is more likely to be in the region of two to four percent per annum over the foreseeable future. Having said that, demand for flexible packaging in Eastern Europe is currently increasing faster than Western Europe, with Poland achieving in excess of five percent growth.

According to a recent PCI Wood Mackenzie report, Europe is now one of the biggest and most sophisticated flexible packaging markets in the world, and accounted for around 17 percent of the global total in 2017. However, it is becoming increasingly mature, although expected to remain a world leader in technical innovation and know-how, as well as being a major exporter.

In terms of market value, current forecasts for 2018 indicate that the value of the global flexible packaging market was around 170 billion USD in 2017, and rising to between 225 and 300 billion USD by 2024. A few industry forecasters are predicting even higher.

Variations in market size and growth forecasts in part depend on the particular reference sources used: flexible packaging may be shown in some studies as including both shrink and stretch wrap, shrink sleeve and stretch sleeve applications. However, they all seem to agree that the largest market for all the different types of flexible packaging solutions can be found in food (Figure 7.1) and beverages (both retail and institutional), which accounts for a near 60 percent of all shipments (and nearer to 80 percent in some markets). As a guide, the key end-use markets for flexible packaging are featured in Figure 7.2.

As can be identified from the chart, a key part of the upsurge in demand for flexible packaging has come from (packaged and processed) food applications, with changing lifestyle and eating habits all having a positive influence on the overall market share emanating from food and beverages applications. Such changes are in part driven by innovations in food processing and packaging that contribute to increasing the shelf life of foods – led by a whole range of snack and convenience foods, ready prepared meals, coffee pouches and on-the-go food and beverages – as well as the proliferation of quick-serve restaurants and cafés and changes in retail food trading.

Other important drivers in the growth of flexible packaging include:

- Continued rapid growth in the global pharmaceutical and medical industry

Figure 7.1 The largest market for flexible packaging is in food products and food snacks

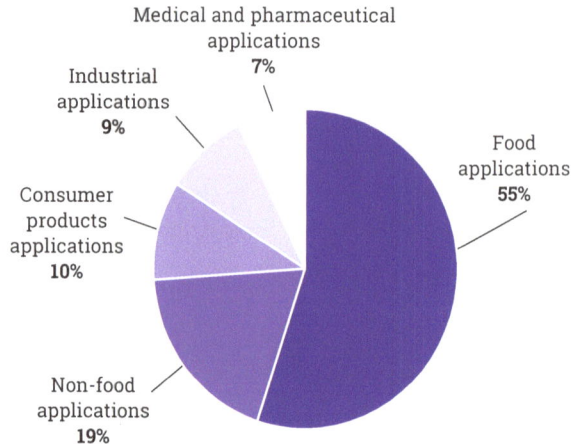

Figure 7.2 Global flexible packaging market – the key end-use markets for flexible packaging

- New and changing applications in the health and personal care sectors
- Continuing growth in the market for cosmetics products
- High potential growth in retail non-food applications
- Growth opportunities in household and consumer products
- New growth opportunities in the nutritional and supplements markets
- Continued growth for agricultural, horticultural and veterinary products
- The potential in pet food applications
- Other new applications, including sporting goods.

MARKET OPPORTUNITIES FOR NARROW- AND MID-WEB PRINTING

For the label converter looking to move into flexible packaging it is interesting to note that all the key end-use markets and the drivers of growth are virtually the same for both labels and flexibles. The label converter will often already be supplying labels to the same customers who are sourcing flexibles.

Flexible packaging buyers are certainly now looking for more cost-effective (or added-value) solutions, faster job turnarounds, versioning and variable data, just-in-time and on-demand printing, and reduced inventory costs. Something that the self-adhesive label converter has long been called upon to provide to his customers.

This all means that with flexible packaging run lengths and pack sizes getting smaller, as well as the possibility of adding in-line value and perhaps personalization, multi-versions and variations, there are undoubted opportunities for narrower – and particularly mid-web – label converters (using both flexo and digital technology) to capture a growing portion of the flexibles market that is not suited to wider web CI flexo or gravure presses as they struggle to print the smaller and higher added-value print and finishing run lengths economically.

Indeed, research in the US has already shown that jobs with the shortest run sizes are in pouches of any kind, as well as single serve and one-dose packs and sachets. These are key areas where label converters can most readily enter the flexible packaging market. Narrower and mid-web (especially in the 430-450 mm range) conventional and digital printing has effectively opened up the market to a whole range of new customers who have never previously had a solution for short-run flexible packaging orders.

Traditionally, lead times for flexible packaging printing on wide format flexo and gravure presses have been anything up to 40 or more days. Today, narrower and mid-web flexo and digital production

Figure 7.3 Small run lengths, multiple SKUs, on demand printing and fast turnaround on HP Indigo 20000. Source: ePac

FLEXIBILITY OF DIGITAL

With more and more installations around the world, many label converters are also now seeing value in the production of flexible packaging using digital printing. Although still only making up about one to two percent of total global flexible packaging volume, there is definitely an ongoing shift towards digital. Digital printing's ability and flexibility to react swiftly to market demands – and to produce small print and custom runs, multiple SKUs (see Figure 7.3), on demand, without the corresponding loss of time that accompanies the production of the new tooling required for flexo, offset or rotogravure printing – offers significant and positive benefits. Digital printing, whether using toner or inkjet, effectively complements the traditional analogue methods.

Growth expectations for all formats of digital printing are currently around 15 percent per year – considerably higher than conventional analogue printing. With digital packaging platforms opening up possibilities for creating new revenue streams, everyone in print packaging has an opportunity to take advantage of the latest digital and hybrid technologies which are now giving brands and their design teams more chances to use their creativity in new ways. Exciting for them and potentially lucrative for narrow- and mid-web label and flexible packaging companies to position themselves to benefit from these advances.

Digital printing technology is already being more widely used in the industry to serve the purpose of labeling/decorating flexible packaging. Enhanced safety, less maintenance, efficient energy consumption and minimizing waste are some of the beneficial factors enhancing digital printing.

Narrower and mid-web label converters can also use digital print and in-line finishing to provide hard to replicate brand protection features on packs and labels, aiding anti-counterfeit efforts. The capability of digital systems to impart variable information – sequential coding or numbering, or an item-specific QR code – onto a pack or label also provides new security or traceability capabilities.

To this list of these more specific market sectors opportunities for the narrow- and mid-web converter can be added other more general changes in the market place that have an impact on flexible packaging growth opportunities, such as:

technology has been able to bring this down to nearer ten days, or even less.

Certainly, there are many narrower and mid-web in-line style printing presses (servo-driven machines in narrower web widths) today that are now able to handle these shorter flexible packaging pouch and sachet runs that fit within the press web widths. Servo drives and controls have helped overcome the main challenges that have historically plagued the flexo market, such as gear marking, while ever-more sophisticated web handling and temperature control features on narrower web presses have made the handling and printing of thinner films far more accurate and precise.

Perhaps not unsurprisingly, a recent study of label converters in Europe found that 19 percent of them were already involved to some degree in the production of flexible packaging, while over 40 percent of visitors to Labelexpo Europe 2017 said that they had some responsibility for flexible packaging.

A similar study in North America found that narrower web label converters that had entered the US flexible packaging market were achieving an average of 9.73 percent per annum growth, with some of the leading narrow- and mid-web producers obtaining a growth rate of as high as 15 percent per annum growth. Certainly, better rates of growth than many of the more traditional label converters are currently achieving.

- Ever-increasing demand for packaged foods
- Shrinking household sizes (ready meals for one or two people)
- A growing preference for flexible packaging over rigid packaging
- Extended retail shelf-life requirements
- Rising disposable incomes
- Measures to reduce the carbon footprint
- Changing lifestyles and convenience eating

THE ROLE OF FILMIC MATERIALS

While there are a range of flexible packaging materials and constructions that are available and widely used, it is plastics (polymer) films which now dominate the market and are said to currently account for over 70 percent of the industry's revenue, making them a key area for the narrower and mid-web converter to understand.

Certainly the introduction of ever-more filmic materials since the 1960s, as well as advances in coatings, surface treatments and seals, sachets and pouches, the range of applications and markets for flexible packaging has grown dramatically and now includes packaging for very many different and varied markets.

The role and importance of plastics films was discussed in some detail in Chapter 2 where it was noted that the primary reason for the increasing popularity and growth of plastics in flexible packaging is the highly versatile nature of a wide variety of polymer films now available, which enables them to be converted into a large number of shapes, sizes, and designs, with a whole portfolio of performance characteristics. Also, plastics are more flexible, durable, and cost-effective than many other materials used for flexible packaging which has led to their increased adoption.

However, this does mean that one of the first challenges that the narrow- and mid-web converter will be faced with when moving into flexible packaging is deciding what paper, film, film construction, metalized film, etc., to use or include in a flexible packaging structure with barrier performance. Plenty of polymers, barrier materials, coatings and sealants are available and in common use in the industry. Gaining a basic understanding of the key material choices available, the requirements of different product types and the commercial and technical considerations can undoubtedly simplify the selection process and pay dividends for those entering the market

The aim should be to ask the right questions and agree the key specification requirements with each customer – not only what matters, but also what is known to work. To a large degree this will come down to experience and the implementation of due diligence procedures. The correct substrate specifications and selection will undoubtedly ensure a happy customer. Poor or wrong substrate selection can adversely affect the supply relationship, and may result in high wastage, rejected goods, higher costs and even non-payment.

It is therefore more than worthwhile for label converters to spend some time learning about flexible packaging films, film and barrier constructions, adhesives and laminating if they are looking to achieve a successful operation in this new market area.

FLEXIBLE PACKAGING APPLICATIONS

Previous chapters in this book have already mentioned the main market sectors for flexible packaging: food and beverages, consumer and household products, pharmaceutical and medical, healthcare, cosmetics and toiletries, retail non-food, industrial applications, nutritional and supplements, agricultural, horticulture and veterinary and pet foods.

Most of these sectors are already well served by the narrow- and mid-web label converter, so will be well-known to them in terms of market requirements, lead times, service, quality performance and, probably, with many of the same customers or buyers also sourcing flexible packaging. For the label converter therefore, it maybe is not so much about finding new customers, but about understanding and servicing the shorter-run, smaller-size, added-value and niche flexible packaging requirements of existing customers – something which a growing numbers of label converters have been successfully and profitably achieving in recent years.

To aid the label converter in understanding some of the key flexible packaging market potential sectors and opportunities, the remainder of this Chapter takes a more detailed look at what it might be feasible for them to achieve.

FOOD AND BEVERAGE

Food and beverage flexible packaging, both retail and institutional, is said to account for a near 60 percent of all flexible packaging applications, with some market studies claiming a market share as high as 80

Figure 7.4 Examples from Mexico of pouches/bags used for snack food packaging. Source: Tory Plastics

Figure 7.6 Coffee packaging in a gusseted stand-up pouch. Source: Constantia Flexibles

Figure 7.5 Coffee packaging in sticks. Source: Constantia Flexibles

percent. It largely depends on exactly what a specific study includes. Does it include shrink sleeves or not? Does it include industrial applications?

Whichever way the market is analyzed, food and beverages are by far the biggest markets for flexible packaging. Indeed, as they are for labels. They therefore provide the biggest opportunity for the narrow- and mid-web converter.

In particular, the rising demand for snack and convenience foods in small or single size portions (Figure 7.4), increased spending on bakery and cereal bars, short-run ready meals and coffee or hot chocolate sticks (see Figure 7.5) and pouches, dehydrated and dry foods (instant soup, gravy and sauce packets, rice, food mixes), snack foods (again Figure 7.4) and nuts, spice foods, chocolates and

sweets, ice-cream novelties, bakery products such as cookies (biscuits), cakes and chips (crisps).

Stand up gusseted pouches (Figure 7.6), and pillow pouches are widely used as flexible packaging products in the food sector. Pillow pouches have also witnessed high gains due to their increasing usage in the food, beverages and dairy industries. Low cost, high sealing ability and cost efficient transportation are some of the key properties positively influencing product penetration.

Stand-up pouches have become one of the most preferred flexible packaging products, owing to their versatility in various end-use food applications. Stand-up pouches are estimated to account for a revenue share of 75 billion USD by 2024 owing to attributing features that include high barrier properties against oxygen and moisture, low material consumption, and cost-effective properties.

Thermally processed foods are packed in these products to extend the shelf life. Other flexible packaging products used in the food sector include retort pouches, four side seal pouches, a wide variety of lidding, and portion packs. Lidding is also widely used for yogurt, cream and dessert pots.

A typical lidding film application is illustrated in Figure 7.7. Lidding films not only seal and protect the product but can perform an important decorative function. Most lidding films are designed to be peelable to allow easy access to pack contents

Pre-printed wrappers and decorative bands are another effective way for flexible pack branding and packaging of foodstuffs. Flow wrappings are typically

Figure 7.7 A ready meal lidding film application. Source: OPM Group

Figure 7.8 Illustration shows examples of flow wrapped products

used on confectionery, sweets, butter packs, etc. In many cases products are completely over wrapped in paper or film to form an airtight seal. Figure 7.8 is a good example of this style of decorative packaging.

A steadily shifting preference for aesthetically appealing food products is also fuelling flexible packaging market demand. Increasing extended shelf life requirements along with preservation from contamination also open new avenues for industry growth in the food and beverages sector, along with rising consumer consciousness pertaining to food safety and hygiene.

HEALTH, BEAUTY AND PERSONAL CARE

The healthcare, cosmetics, toiletries and personal care markets have long provided a wealth of opportunities for the label converter. Now, they also offer the opportunity to diversify into the production of high quality sachets (Figure 7.9), pouches and packs for many different – often shorter run lengths of multiple versions and variations – types of flexible packaging for the whole health and personal care market that includes toiletries, hygiene, shampoo, liquid soaps, creams, lotions, gels, cosmetics products, beauty products, wipes and packs.

Flexible packaging technology is one of the key sectors predicted to witness significant growth over the coming years owing to growing middle-class populations, rising disposable incomes, and the escalating demand for all types of healthcare, cosmetics and toiletries products. New product

Figure 7.9 Flexible packaging wraps and sachets used for beauty and personal care. Source: OPM Group

launches by major brands in emerging markets (such as in India and China) are also projected to motivate the growth of the global cosmetics sector.

Ongoing innovations and applications in this sector

Figure 7.10 Single-use supplement and nutritional pack. Source: OPM Group

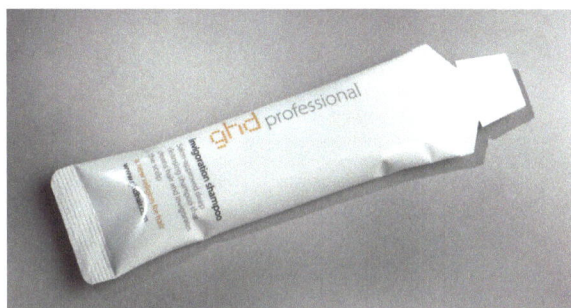

Figure 7.11 Stickpack packaging used for nutritional, pharmaceutical and haircare products. Source: OPM Group

which are predicted to continue to drive growth opportunities for label converters include all kinds of creams and gels (used extensively by the cosmetics, personal care and haircare industries), impregnated hand and tissue wipes, nutricosmetics, travel-sized creams, bath soaks, non-free-flow powders such as talcum powder – many of these applications being in sachets.

PHARMACEUTICAL, NUTRITIONAL, NUTRACEUTICAL, VETERINARY AND MEDICAL

The pharmaceutical and related sectors are variously recorded as the second biggest application section (around nine to 10 percent) for flexible packaging, with both trends in healthy living and increasing incidences of lifestyle related diseases among the working population expected to boost the demand for pharmaceutical, nutraceutical and medical items – which in turn will optimistically affect the flexible packaging and narrower web markets of tomorrow.

This sector is quite diverse in its array of products, which today ranges through all kinds of pharmaceutical products and ointments to weight loss and dietary supplements, sports nutrition supplements such as protein drink mixes and shakes, pre-workout powders, energy bars, foods for special dietary needs, powdered energy or vitamin drinks, spices, effervescent tablets, capsules and powders.

Single-dose packaging such as single-use pouches (Figure 7.10) and stickpacks provide today's on-the-go consumers of nutritional and nutraceutical products with portability and convenience.

Stickpack packaging (Figure 7.11) in particular, is widely used by many leading nutritional brands for products such as drink mix supplements and protein powders, especially offered with easy open or tear propagation or laser scoring.

ANIMAL CARE

It was interesting to note in the recent PCI Wood Mackenzie study of flexible packaging in Europe that they identified pet food as being the best performing flexible packaging category, currently growing in Europe at around four percent per annum.

A similar study in the US by the Freedonia Group pegged the US pet food packaging market at about 2.5 billion USD and again forecasts growth at around four percent. Pouches for pet foods (Figure 7.12) and pet products are identified as the fastest growing segment of the animal care market, due to them being easy to carry, store and re-seal to ensure freshness.

Within the whole animal care sector are a wide range of products suitable for printing by label converters that include pet foods, pet care products, pet treats, veterinary powders and animal products, wild bird and wild animal feeds – with a great many short-runs of single serve pouches and gusset bags being produced on a regular basis for wet, semi-wet and dry pet foods.

Pet food packaging can be a challenge. It needs

Figure 7.12 Pet and animal products. Source: OPM Group

to prevent spillage, stand-up to multiple use, and maybe offer convenience features, such as re-closure or handle options.

RETAIL NON-FOOD, HOME AND GARDEN
Depending on the particular market study viewed, the whole non-food, home, DIY and garden sector accounts for up to 10 percent of the flexible packaging market, with products that include household items, laundry detergents, soaps, bath salts, under-the-sink packs, paints, pastes and plaster, lawn care, fertilizer, pesticides, grass seed, compost and bark.

Many of these products are packed in large bags and not suitable for the narrow- and mid-web converter, but smaller-sized flexible pouches and bags – both stand-up and flat – are also widely used and do provide an opportunity for single-use situations, such as flower and vegetable seeds.

INDUSTRIAL APPLICATIONS
With some studies showing industrial applications for flexible packaging being as high as six percent of the market, there are undoubtedly some opportunities for the label converter, especially for label companies that are already supplying labels into this sector and have potential customers that would also buy stand-up pouches and bags or sachet products from them.

Flexible packaging for industrial applications may require custom-made bag and pouch films and laminated rollstock specifically constructed to handle industrial powders, granular chemicals, lubricants, agricultural products and other such applications.

MAKING THE MOST OF THE OPPORTUNITIES
As a number of label converters are already proving, flexible package printing represents an opportunity for PS label converters to widen their product portfolio, compete on short-runs and smaller packs, and utilize high quality flexo, digital, combination and hybrid press technology (plus in-line lamination, added value embellishing or cold foiling), to increase the range and variety of high quality printed products produced.

Put together, the right width narrow- and mid-web presses, both conventional and digital, now offer label and flexible packaging converters the opportunity to target new short-run flexible packaging applications, multiple SKUs, to drive differentiation and personalization, offer faster turnaround and quicker delivery and reduced stockholding – and become more profitable.

As can be seen, the possible opportunities cover almost all types of consumer food, pharmaceutical, health and beauty, nutritional, garden, DIY, leisure and other retail market applications, as well as increasingly moving in to the industrial, automotive, agricultural, horticultural and medical sectors. Packaging types open to the label converter for this variety of products and applications are largely found or seen as being in stand-up pouches, lay flat pouches, sachets, lidding, and roll flow-wrapping materials.

But it's not just about the print opportunities, successful converters targeting the flexible packaging market are particularly finding new business opportunities with the independent and more regional brands, helping these smaller and medium-sized company buyers to simplify how they buy their flexible packaging, and guiding them in understanding the specifications, tolerances, materials, origination, color technology, inks, and print requirements.

Quite simply, the more successful narrower and mid-web flexible packaging converters today are creating a new type of service model that makes the selection and purchasing of high quality flexible packaging in smaller sizes and shorter runs as fast and easy as possible.

Index

www.ingramcontent.com/pod-product-compliance
Lightning Source LLC
Chambersburg PA
CBHW041721210326
41598CB00007B/739